Dennis Simon

Aspects in the fate of primordial vacuum bubbles

Dennis Simon

Aspects in the fate of primordial vacuum bubbles

Südwestdeutscher Verlag für Hochschulschriften

Impressum/Imprint (nur für Deutschland/only for Germany)
Bibliografische Information der Deutschen Nationalbibliothek: Die Deutsche Nationalbibliothek verzeichnet diese Publikation in der Deutschen Nationalbibliografie; detaillierte bibliografische Daten sind im Internet über http://dnb.d-nb.de abrufbar.
Alle in diesem Buch genannten Marken und Produktnamen unterliegen warenzeichen-, marken- oder patentrechtlichem Schutz bzw. sind Warenzeichen oder eingetragene Warenzeichen der jeweiligen Inhaber. Die Wiedergabe von Marken, Produktnamen, Gebrauchsnamen, Handelsnamen, Warenbezeichnungen u.s.w. in diesem Werk berechtigt auch ohne besondere Kennzeichnung nicht zu der Annahme, dass solche Namen im Sinne der Warenzeichen- und Markenschutzgesetzgebung als frei zu betrachten wären und daher von jedermann benutzt werden dürften.

Coverbild: www.ingimage.com

Verlag: Südwestdeutscher Verlag für Hochschulschriften GmbH & Co. KG
Heinrich-Böcking-Str. 6-8, 66121 Saarbrücken, Deutschland
Telefon +49 681 37 20 271-1, Telefax +49 681 37 20 271-0
Email: info@svh-verlag.de

Approved by: Würzburg, Julius-Maximilians-Universität, Diss., 2012

Herstellung in Deutschland:
Schaltungsdienst Lange o.H.G., Berlin
Books on Demand GmbH, Norderstedt
Reha GmbH, Saarbrücken
Amazon Distribution GmbH, Leipzig
ISBN: 978-3-8381-3154-2

Imprint (only for USA, GB)
Bibliographic information published by the Deutsche Nationalbibliothek: The Deutsche Nationalbibliothek lists this publication in the Deutsche Nationalbibliografie; detailed bibliographic data are available in the Internet at http://dnb.d-nb.de.
Any brand names and product names mentioned in this book are subject to trademark, brand or patent protection and are trademarks or registered trademarks of their respective holders. The use of brand names, product names, common names, trade names, product descriptions etc. even without a particular marking in this works is in no way to be construed to mean that such names may be regarded as unrestricted in respect of trademark and brand protection legislation and could thus be used by anyone.

Cover image: www.ingimage.com

Publisher: Südwestdeutscher Verlag für Hochschulschriften GmbH & Co. KG
Heinrich-Böcking-Str. 6-8, 66121 Saarbrücken, Germany
Phone +49 681 37 20 271-1, Fax +49 681 37 20 271-0
Email: info@svh-verlag.de

Printed in the U.S.A.
Printed in the U.K. by (see last page)
ISBN: 978-3-8381-3154-2

Copyright © 2012 by the author and Südwestdeutscher Verlag für Hochschulschriften GmbH & Co. KG and licensors
All rights reserved. Saarbrücken 2012

Zusammenfassung

Die Idee der kosmologischen Inflation stellt heute die wichtigste Erweiterung der klassischen Urknalltheorie dar. Seit ihrem Aufkommen in den frühen 80er Jahren sind zahlreiche physikalische Mechanismen bekannt und ausgearbeitet geworden, die die inflationäre Expansion des Raums vor der dem 'heißen' Urknall auf eine tragfähige, theoretische Basis stellen. Zu den Errungenschaften der Inflationstheorie zählen unter Anderem die Erklärung der nahezu Euklidischen Geometrie des sichtbaren Raums, die bemerkenswerte Homogenität der kosmischen Hintergrundstrahlung, im Besonderen aber auch die ihr innewohnenden winzigen Unregelmäßigkeiten mit einer relativen Amplitude der Größenordnung 10^{-5}. In vielen Inflationsmodellen endet die Inflation allerdings nur lokal. Demzufolge besteht die Möglichkeit, dass es außerhalb des von uns sichtbaren Raums Gebiete geben kann, in denen der inflationäre Prozess weiterhin stattfindet. Dieser Eigenschaft wird durch den Begriff 'Ewige Inflation' Rechnung getragen.

Ewige Inflation kann sich im Rahmen der Skalarfeld-Inflation in verschiedenen Formen manifestieren. Zum Einen können die Fluktuationen des Feldes so groß sein, dass sie der klassischen Trajektorie, und damit dem Ende der Inflation, entgegenwirken wirken. In Regionen, in denen das geschieht, setzt sich die beschleunigte Expansion des Raums mit einer höheren Rate weiter fort. In Teilen solcher Regionen mag sich dies wiederholen und der Vorgang auf diese Weise theoretisch bis ins Unendliche andauern. Raum und Feld reproduzieren sich selbst. Eine weitere Möglichkeit, die sowohl unabhängig als auch zusätzlich zur zuvor beschriebenen auftreten kann, ist die des sogenannten Vakuumtunnelns. Wenn das Potential des Skalarfelds mehrere lokale Minima aufweist, so legt eine semi-klassische Rechnung, dass das Feld innerhalb eines sphärischen Gebiets, einer Blase, in einen anderen Zustand tunneln kann. Die zugehörigen Tunnelraten hängen von der Differenz der beiden Minima und der Gestalt des Potentials zwischen ihnen ab. Gewöhnlich ist diese Rate exponentiell unterdrückt, was bedeutet, dass die Inflation sehr lange andauert bevor es zu einem Tunnelprozess kommt. Fortwährende Inflation beseitigt großräumig, effektiv, jegliche Form der Unregelmäßigkeit, d. h. Raumkrümmung, Anisotropie und Inhomogenität, sodass dieser Sachverhalt unter dem Ausdruck 'cosmic no-hair conjecture' bekannt ist. Aus diesem Grund waren bisherige Betrachtungen fast ausschließlich der Entwicklung von Blasen in einem Vakuumhintergrund (der de Sitter Raumzeit) gewidmet. Neue Überlegungen im Rahmen der Stringtheorie erlauben allerdings auch hohe Tunnelraten, sodass die Möglichkeit der Nukleation von Blasen in nicht-vakuumdominierten Hintergründen besteht. Die weitere Entwicklung hängt in diesem Fall von den Eigenschaften des Hintergrunds ab. Diese Tatsache ist Gegenstand des ersten Teils dieser Arbeit.

Einer vertiefenden Einführung in Kapitel 4 folgt die Vorstellung der Lemaître-Tolman Raumzeit in Kapitel 5, die die Hintergrundraumzeit in der Studie von Auswirkungen der Inhomogenität auf die Entwicklung von Vakuumblasen bildet. In Kapitel 6 wird der 'thin-shell' Formalismus und die sich daraus ergebenden Gleichungen explizit dargelegt. Dem schließt sich in Kapitel 7 eine detaillierte Analyse der Blasenentwicklung in verschiedenen Limites des Lemaître-Tolman Hintergrunds und in einer Robertson-Walker Raumzeit mit rapidem Phasenübergang an. Nach der Ableitung der Vakuumlösung wird sukzessive auf die Blasenentwicklung in einem statischen Hintergrund, in einem dynamischen, aber homogenen Hintergrund, in einem flachen, inhomogenen Hintergrund, in einem gekrümmten, inhomogenen Hintergrund und in einem homogenen Hintegrund mit Phasenübergang eingegangen. Zu den zentralen Beobachtungen gehört, dass die Präsenz des Staubs, bei fixierter Oberflächendichte, eine Verringerung des Nukleationsvolumens mit sich bringt und dazu führen kann, dass die Blase einen Kollaps beginnt. Darüber hinaus zeigt sich, dass in einem expandierenden Hintergrund ein radial inhomogenes Staubprofil hinreichend schnell verdünnt wird, sodass praktisch kein Effekt auf die Blasenwand zu erkennen ist. Das ändert sich in einem radial inhomogenen Krümmungsprofil, positive Raumkrümmung hat einen abbremsenden Effekt auf die Expansion der Blase. Weiterhin erkennt man, dass eine Änderung der Zustandsgleichung von $w = -1$ zu $w = 1/3$ in einem homogenen Flüssigkeitshintergrund ebenso zu einer Abbremsung der Blase führt. Es wird herausgestellt, dass der verwendete Ansatz keine Möglichkeit zur Behandlung eines, physikalisch zu erwartenden, Materietransfers bietet und die damit erzielten Ergebnisse unter diesem Vorbehalt zu verstehen sind.

Im zweiten Teil der vorliegenden Arbeit wird potentiell beobachtbaren Konsequenzen der Kollision zweier Blasen in der kosmischen Hintergrundstrahlung nachgegangen.

Die topologische Natur des Signals in der letzten Streufläche legt die Verwendung von Statistiken nahe, die es erlauben, die morphologischen Eigenschaften der Temperaturfluktuationen zu quantifizieren. Diese Statistiken bieten die Minkowski Funktionale (MF), die in Kapitel 10 vorgestellt werden. Dem schließt sich in Kapitel 11 die Präsentation des den Simulationen zu Grunde liegenden Softwarepaketes an. Dieses wird benutzt um Karten eines Gaussschen Zufallsfeldes zu erzeugen und die entsprechenden MF zu berechnen. Die vorgestellte Fehleranalyse erlaubt eine höhere Präzision der numerischen MF im Vergleich zu bisherigen Methoden. Im folgenden Kapitel 12 präsentieren wir die Anwendung des Algorithmus' auf Gausssche- und Kollisionskarten. Die zu erwartenden MF werden motiviert, numerisch erfasst und verglichen. Ein Fit der geringsten quadratischen Abweichung reproduziert die tatsächlichen Parameter nur dann, wenn über eine hohe Anzahl von Realisierungen gemittelt wird, wohingegen die Betrachtung einer einzigen Karte in WMAP bzw. Planck Auflösung nur für "auffällige" Scheiben mit Temperaturunterschied $\delta T \gtrsim 2\sqrt{\sigma_G}$ und Öffnungswinkel $\vartheta_d \gtrsim 40°$ Übereinstimmung erreicht wird. Dies bedeutet, das MFe ein schlechtes 'Signal zu Rausch'Verhältnis besitzen um heiße oder kalte Scheiben in der kosmischen Hintergrundstrahlung zu erfassen. Diese Vermutung wird weiter gestärkt durch die Tatsache, dass die Anwendung der Methode auf die WMAP7 Daten kein konsistentes Resultat liefert.

Den beiden beschriebenen Teilen geht noch eine allgemeine, aber elementare Einführung in das Themengebiet voraus. Material, das nicht unmittelbar zum Verständnis der Arbeit nötig ist, an manchen Stellen aber unterstützend sein könnte, ist in einem Appendix untergebracht.

Abstract

At the present day the idea of cosmological inflation constitutes an important extension of Big Bang theory. Since its appearance in the early 1980's many physical mechanisms have been worked out that put the inflationary expansion of space that proceeds the Hot Big Bang on a sound theoretical basis. Among the achievements of the theory of inflation are the explanation of the almost Euclidean geometry of 'visible'space, the homogeneity of the cosmic background radiation but, in particular, also the tiny inhomogeneity of a relative amplitude of 10^{-5}. In many models of inflation the inflationary phase ends only locally. Hence, there exists the possibility that the inflationary process still goes on in regions beyond our visual horizon. This property is commonly termed 'eternal inflation'.

In the framework of a cosmological scalar fields, eternal inflation can manifest itself in a variety of ways. On the one hand fluctuations of the field, if sufficiently large, can work against the classical trajectory and therefore counteract the end of inflation. In regions where this is the case the accelerated expansion of space continues at a higher rate. In parts of this region the process may replicate itself again and in this way may continue throughout all of time. Space and field are said to reproduce themselves. On the other hand, a mechanism that can occur in addition or independent of the latter, is so called vacuum tunneling. If the potential of the scalar field has several local minima, a semi-classical calculation suggests that within a spherical region, a bubble, the field can tunnel to another state. The respective tunneling rates depend on the potential difference and the shape of the potential between the states. Generally, the tunneling rate is exponentially suppressed, which means that the inflation lasts for a long time before tunneling takes place. The ongoing inflationary process effectively reduces local curvature, anistotropy and inhomogeneity, so that this property is known as the 'cosmic no-hair conjecture'. For this reason cosmological considerations of the evolution of bubbles thus far almost entirely involved vacuum (de Sitter) backgrounds. However, new insights in the framework of string theory suggest high tunneling rates which allow for the possibility of bubble nucleation in non-vacuum dominated backgrounds. In this case the evolution of the bubble depends on the properties of the background spacetime.

This phenomenon is the subject of the first part of this thesis. A deeper introduction in chapter 4 is followed by the presentation of the Lemaître-Tolman spacetime in chapter 5 which constitutes the background spacetime in the study of the effect of matter and inhomogeneity on the evolution of vacuum bubbles. In chapter 6 we explicitly describe the application of the 'thin-shell' formalism and the resulting system of equations. This is succeeded in chapter 7 by the detailed analysis of bubble evolution in various limits of the Lemaître-Tolman spacetime and a Robertson-Walker spacetime with a rapid phase transition. After a review of the vacuum solution we consider in particular the bubble evolution in a static background, in a homogeneous dynamic background, in a flat inhomogeneous background and in an homogeneous background with phase transition. The central observations are that the presence of dust, at a fixed surface energy density, goes along with a smaller nucleation volume and possibly leads to a a collapse of the bubble. In an expanding background, the radially inhomogeneous dust profile is efficiently diluted so that there is essentially no effect on the evolution of the domain wall. This changes in a radially inhomogeneous curvature profile, positive curvature decelerates the expansion of the bubble. We observe that a change in the equation of state parameter from $w = -1$ to $w = 1/3$ in a homogeneous fluid background also leads to a deceleration of the bubbles expansion. Moreover, we point out that the adopted approach does not allow for a treatment of a, physically expected, matter transfer so that the results are to be understood as preliminary under this caveat.

In the second part of this thesis we consider potential observable consequences of bubble collisions in the cosmic microwave background radiation.

The topological nature of the signal suggests the use of statistics that are well suited to quantify the morphological properties of the temperature fluctuations. In chapter 10 we present Minkowski Functionals (MFs) that exactly provide such statistics. This is followed by chapter 11 where we present the software package that is used for the production Gaussian maps and for the extraction of MFs which builds the basis for further simulations. The presented error analysis allows for a higher precision of numerical MFs in comparison to earlier methods. In chapter 12 we present the application of our algorithm to a Gaussian and a collision map. We motivate the expected MFs and extract their numerical counterparts. We find that our least-squares fitting procedure accurately reproduces an underlying signal only when a large number of realizations of maps are averaged over, while for a single WMAP and PLANCK resolution map, only when a highly prominent disk, with $|\delta T| \gtrsim 2\sqrt{\sigma_G}$ and $\vartheta_d \gtrsim 40°$, we are able to recover the result. This is unfortunate, as it means that MF are intrinsically too noisy to be able to distinguish cold and hot spots in the CMB for small sizes. In order to confirm our suspicion, we have applied our prescription to WMAP7 map and found that we did not recover the latter's conclusions.

The two described parts are preceeded by a general but elementary introduction to the field. Material that supplements the text, though not of immediate necessity for comprehension, is relegated to an appendix.

Per aspera ad astra

Contents

Introduction   3

1. Outline of standard cosmology   3
     1.1. Friedmann-Lemaître-Robertson-Walker cosmology 3
     1.2. Present day values of the expansion rate and energy density parameters . 4
     1.3. Current view of the evolution of matter and space 6
     1.4. Discomfort with the standard Big Bang scenario 8

2. The paradigm of cosmological inflation   9
     2.1. Scalar field inflation 9
     2.2. Debate on initial conditions 11

3. Eternal inflation, vacuum decay and bubble collisions   13
     3.1. Eternal inflation 13
     3.2. Decay of a metastable vacuum 15
     3.3. Classical bubble evolution 16
     3.4. Observability of bubble collisions 18

I. Vacuum Bubbles on Dynamical Backgrounds   19

4. Introduction to Part I   21

5. The Lemaître-Tolman spacetime   23

6. Spacetime junction   27
     6.1. General considerations 27
     6.2. Explicit choice of the junction geometry 29
     6.3. System of equations to be solved 30

7. Evolution on dynamical backgrounds   33
     7.1. Evolution on vacuum background 33
     7.2. Evolution in homogeneous dust 35
     7.3. Evolution through inhomogeneous dust 41
     7.4. Evolution during a rapid phase transition 44

8. Summary and conclusion of Part I   47

Contents

II. Signatures of Bubble Collisions in Minkowski Functionals   49

9. Introduction to Part II   51

10. Minkowski Functionals on the two-sphere   53
10.1. Definition 53

11. Application of the HEALPix software   55
11.1. Basic introduction to HEALPix 55
11.2. Extraction of Minkowski Functionals 56

12. Minkowski Functional statistics of a collision signal   59
12.1. Gaussian random field 59
12.2. Analysis of collision maps 66

13. Summary and conclusion of Part II   73

Appendix   77

A. Spherically symmetric spacetime   77

B. Lemaître-Tolman spacetime   79

C. Foliations of de Sitter spacetime   81

D. Relativistic Junction Conditions   83

E. Minkowski functionals on the sphere   85
E.1. Transformation to surface integrals 85
E.2. Gaussian Random fields 86

Bibliography   95

Introduction

1. Outline of standard cosmology

1.1. Friedmann-Lemaître-Robertson-Walker cosmology

Several basic features of modern cosmology are well established today. Together, progress in theory and precise observations have revealed the huge scale $\approx 3000\text{Mpc}$ of the observable universe as well as the fact that it is evolving and expanding. Physical cosmology relies on the assumption that the laws of physics are the same in all of spacetime. The only relevant force on astronomical scales is gravity and consequently standard cosmology is based on the classical relativistic theory of gravitation, Einstein's General Theory of Relativity. In General Relativity the properties of the matter distribution as expressed in the stress-energy tensor $T_{\mu\nu}$ are the source of spacetime curvature according to Einstein's field equations

$$G_{\mu\nu} + \Lambda g_{\mu\nu} = 8\pi T_{\mu\nu}\,.$$

The Einstein tensor $G_{\mu\nu} := R_{\mu\nu} - g_{\mu\nu}\mathcal{R}/2$ is related to the Ricci tensor $R_{\mu\nu}$ and Ricci scalar $\mathcal{R}$ which contain second derivatives of the spacetime metric $g_{\mu\nu}$ and are non linear. The constant Λ is called the cosmological constant and could in principle also be included in the stress-energy tensor.

Standard cosmological theory adopts the Copernican principle which states that we are not privileged observers. Together with the fact that upon averaging over a large enough spatial scale ($\gtrsim 100\text{Mpc}$) observations become statistically isotropic, this assumption implies spatial homogeneity, cf. Stoeger et al. (1995). The spacetime manifolds that embody spatial homogeneity and isotropy belong to the class of Robertson-Walker geometries, cf. Robertson (1929); Walker (1935).

In a spacetime with Robertson-Walker geometry comoving coordinates, i.e. $u^\mu = \delta^\mu_t$, can be chosen so that the line element becomes

$$ds^2 = -dt^2 + a^2(t)\left(dr^2 + k^{-1}\sin^2\left(\sqrt{k}r\right)d\Omega^2\right)\,.$$

The spatial sections have hyperbolic ($k < 0$), Euclidean ($k = 0$), or spherical ($k > 0$) geometry. Isotropy implies that the stress-energy tensor $T_{\mu\nu}$ necessarily takes the form of a perfect fluid relative to the flow of u_μ, that is

$$T_{\mu\nu} = (\rho + p)\,u_\mu u_\nu + p g_{\mu\nu}\,. \tag{1.1}$$

Here ρ and p are energy density and pressure of the matter distribution and are constant on spatial slices. The flow lines of u^μ represent the wordlines of the matter constituents. The continuity equation $\nabla_\mu T^{\mu\nu} = 0$ turns into

$$\partial_t \rho + 3\frac{\partial_t a}{a}\left(\rho + p\right) = 0\,, \tag{1.2}$$

1. Outline of standard cosmology

The scale factor a obeys the Raychaudhuri equation

$$\frac{\partial_t^2 a}{a} = -\frac{4\pi}{3}(\rho + 3p) + \frac{\Lambda}{3}.$$

Given $\partial_t a \neq 0$, the Raychaudhuri and continuity equation yield the Friedmann equation

$$\left(\frac{\partial_t a}{a}\right)^2 + \frac{k}{a^2} = \frac{8\pi\rho}{3} + \frac{\Lambda}{3}. \quad (1.3)$$

as a first integral. The whole class of solutions to these equations is commonly referred to as Friedmann-Lemaître-Robertson-Walker (FLRW) models. With a determinate matter description, given explicitly or implicitly by an equation of state, there exists a unique solution to the system. In the special case of a linear equation of state $p = w\rho$ the continuity equation (1.2) implies

$$\rho = \rho_0 \left(\frac{a_0}{a}\right)^{3(1+w)}. \quad (1.4)$$

The equation of state parameter of non-relativistic matter is $w = 0$, $w = 1/3$ for relativistic matter and radiation, while a fluid with $w = -1$ is equivalent to a cosmological constant. Accordingly, the scale factor obeys $a \propto t^{2/(3+3w)}$ and $a \propto \exp(Ht)$ respectively.

Initial data that specifies a particular solution at a time t_0 consists of the Hubble rate $H := \partial_t a/a$, the energy density parameters of vacuum Ω_Λ, and the fluid components $\Omega_i := 8\pi\rho_i/(3H^2)$ together with their equation of state. Their sum equals the total energy density $\Omega := \Omega_\Lambda + \sum_i \Omega_i$, and its deviation from unity defines the curvature parameter $\Omega_k := -k/(aH)^2 = 1 - \Omega$. Accordingly, the critical energy density is defined as the energy density at which spatial sections are exactly flat, i.e. $\Omega = 1$.

1.2. Present day values of the expansion rate and energy density parameters

An important role in the determination of the present day Hubbe rate and energy density parameters have measurements of the cosmic microwave background (CMB) e. g. Komatsu et al. (2009a), supernovæ 1A (SNe) by Kowalski et al. (2008), baryon acoustic oscillations (BAO) Percival et al. (2007) and Big Bang nucleosynthesis Fields and Sarkar (2006). The remarkable consistency of these measurements builds a pillar in the so-called concordance model of modern cosmology.

One important cosmological observable is the current expansion rate, or Hubble rate, H_0. First observed by Hubble (1929), several experiments have since confirmed that the light emitted by distant galaxies is redshifted. This observation is interpreted as a relativistic 'Doppler-shift' due to the recession of galaxies by expansion of space. However, this interpretation is misleading, as the distance between comoving observers increases . To account for the uncertainty in the value of the Hubble rate it is customary to use the dimensionless constant h defined through

$$H_0 = 100h \text{ km s}^{-1}\text{Mpc}^{-1}.$$

1.2. Present day values of the expansion rate and energy density parameters

The Hubble radius $H_0^{-1} \simeq 3000 h^{-1}\text{Mpc} \simeq 9.26 h^{-1} 10^{27}\text{cm}$ corresponds to the lengthscale of the presently visible part of the universe, the Hubble time $H_0^{-1} \simeq 9.778 h^{-1} 10^9$ years amounts to the order of magnitude of the time that has passed since the energy density was close to the Planck density, and the critical energy density $\rho_{\text{crit},0} \simeq 1.9 h^2 10^{-29} \text{g cm}^{-3}$.

In the concordance model the value of the Hubble parameter is

$$h \simeq 0.71\,.$$

Moreover, the currently relevant components of the cosmological fluid and their respective equation of state within the concordance model are:

RADIATION: The photons in the universe and any gas of relativistic particles such as neutrinos come with an equation of state $p_\text{r} = \rho_\text{r}/3$. This component contributes only a tiny fraction, $\Omega_{\text{r},0} \approx 10^{-5}$, to the total energy density.

BARYONIC MATTER: The energy density of known species of matter such as atoms and nuclei which are non-relativistic at present represents a fraction of $\Omega_{\text{b},0} \simeq 0.04$ of the total energy density. Being non-relativistic means that their pressure is negligible, $p \ll \rho$, so that their equation of state is approximately $p_\text{b} = 0$.

NONBARYONIC MATTER: The existence of an yet unknown form of matter, commonly referred to as 'Cold Dark Matter' with $\Omega_{\text{c},0} \simeq 0.23$ is required to reconcile several observational facts concerning structure formation and galactic rotation curves, among others. The adjective 'Cold' is meant to express that this matter is also non-relativistic, so that $p_\text{c} = 0$.

DARK ENERGY: The major component in today's cosmic inventory is called 'Dark Energy' as its origin also remains to be understood. The term Dark Energy is associated with any kind of matter whose equation of state is $p_\text{de} \simeq -\rho_\text{de}$. For a cosmological constant this equation holds exactly which makes it a popular candidate. At present, Dark Energy contributes about $\Omega_{\text{de},0} \simeq 0.73$ to the total energy density.

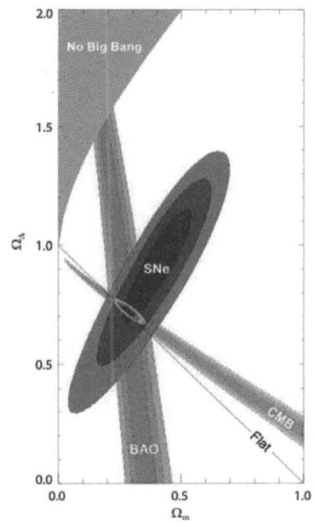

Figure 1.1.: Illustration of the value of present day energy density parameters by a union of different measurements according to Kowalski et al. (2008).

1. Outline of standard cosmology

Consequently, the total energy density is very close to the critical density,

$$\Omega_0 \simeq 1.02 \pm 0.02\,,$$

which means that spatial geometry is almost Euclidean on large scales. The measurement of these parameters allows for a reconstruction of the history of space and matter back to about 13.7 billion years ago as well as for a cautious outlook into their future evolution.

1.3. Current view of the evolution of matter and space

From the Friedmann equation (1.3) and equation (1.4) it follows that the influence of the different components of the cosmological fluid on the expansion of space varies with time. This suggests a classification of cosmological epochs in correspondence with the respective governing source of spatial evolution. The cosmological epochs are:

PRESENT & FUTURE: DARK ENERGY DOMINATION. Today, the evolution of the scale factor has become dominated by Dark Energy. With matter and radiation diluted by the spatial expansion, Dark Energy will eventually remain the single relevant source in the Friedmann equation which will lead to an exponential expansion of space [1].

PAST: MATTER DOMINATED ERA. Roughly six billion years ago, the energy density of matter $\Omega_m = \Omega_b + \Omega_c$ exceeded the energy density of Dark Energy and was the dominant cause of cosmological expansion. This epoch is characterized by the formation of stars and galaxies out of small density perturbations in the cosmological fluid that consists mainly of hydrogen and helium at this time. The era includes the time at which the CMB was released.

REMOTE PAST: RADIATION DOMINATED ERA. Further in the past spatial evolution was dominated by radiation. As temperature decreased a plasma of quarks and gluons combined to baryons and electrons and protons did no longer annihilate. This set the stage for the synthesis of the light elements.

Late in the radiation dominated era temperature and energy density are well beyond the accessible range of particle accelerators. It is possible that there are more particle species than are evident today. For example, according to supersymmetry, the number of particle species must be doubled at least. Moreover, in the classic picture, energy density inevitably increases to the Planck density as we go backwards in time. Quantum effects on geometry can no longer be neglected and General Relativity can no longer be trusted in this case. It is expected that a yet unknown theory of quantum gravity can address physics beyond this scale. The occurrence of infinite energy density and the associated initial singularity within the classic framework is known as the 'Big Bang'.

---

[1] Since the nature of Dark Energy is yet unknown, statements concerning the future evolution must be taken with care.

1.3. Current view of the evolution of matter and space

Figure 1.2.: Temperature fluctuations in the CMB. The statistical properties of these fluctuations allow for a determination of cosmological parameters.

1.3.1. The Cosmic Microwave Backround (CMB) radiation

A very important observable in modern cosmology that is relevant in the second part of this thesis is the cosmic microwave background (CMB) radiation that was first discovered by Penzias and Wilson (1965). Tracing back the evolution of space and matter as described in the last subsection one can infer that the structures we observe today originated from a plasma which once had tiny density perturbations of relative amplitude 10^{-5}. Before ≈ 13.4 billion years ago, the plasma was optically thick, that is photons continously scattered with free electrons so that their mean free path was very short. As the plasma cooled down by spatial expansion nuclei and electrons combined to form neutral atoms and the photons began to free-stream, the associated spatial slice is called the surface of last scattering. Today these photons reach us isotropically with the same temperature distribution as at t_{dec}, an almost perfect blackbody spectrum with central temperature $T = 2.725$K. On their way they were redshifted by cosmic expansion by a factor of $\mathcal{O}\left(10^3\right)$. In the time from the end of inflation (explained in the next chapter) to the surface of last scattering, modes of different wavelengths complete a different number of oscillations leading to a series of maxima and minima in the inhomogeneities on the last scattering surface. Today we measure the power spectrum of anisotropies in the CMB radiation on a two-sphere that is the intersection of our past lightcone with the surface of last scattering.

As remarked in section 1.1 the remarkable homogeneity of the CMB radiation means a good justification for the Robertson-Walker approach to cosmic geometry. Moreover, the spectrum of the density perturbations allows to extract many cosmologically relevant parameters. Recent measurements of the COBE and WMAP collaborations together with more data to be expected from the ongoing Planck mission allow for a determination of these parameters to high accuracy. A detailed presentation of the physics of the cosmic microwave background is to be found in the textbook by Durrer (2008). Despite its remarkable achievements, there is also some discomfort with the Big Bang scenario.

1. Outline of standard cosmology

1.4. Discomfort with the standard Big Bang scenario

Though the standard Big Bang scenario provides a compelling explanation of many cosmological observations, some properties remain that may cause astonishment. Among these are:

SPATIAL FLATNESS. The evolution of the curvature parameter is described by (neglecting Λ)
$$\frac{d\Omega}{d\ln a} = -(1+3w)(1-\Omega)\Omega. \qquad (1.5)$$
Thus, if the equation of state parameter obeys $w > -1/3$, $|1-\Omega|$ will grow with time. The observation that $\Omega_0 \simeq 1.02 \pm 0.02$ implies
$$|\Omega(t_{Pl})| \leq \mathcal{O}\left(10^{-61}\right).$$

HOMOGENEITY ACROSS SUPER HORIZON SCALES. The distance light travels expressed in comoving coordinates is
$$r_p = \int \frac{d\ln a}{aH}.$$
and is called the comoving particle horizon. The hubble radius in comoving coordinates is $(aH)^{-1} = a^{(1+3w)/2} H_0^{-1}$ so that for $w > -1/3$ the comoving particle horizon grows with time. In the the standard big bang scenario the distance light had travelled between $t = t_{Pl}$ and the time of last scattering was released amounts to $\sim 1°$ on the CMB map. Regions separated larger than $1°$ are not in causal contact. The fact that the temperature of the CMB is uniform to 10^{-5} is thus very surprising.

ORIGIN OF STRUCTURE. The structures in the universe are thought to have grown from small density perturbations like those that are observed in the CMB. What is the origin of these perturbations?

Within the standard Big Bang scenario an adequate explanation of these issues can not be given so that they must be anchored within initial conditions. However, space being flat to $\mathcal{O}\left(10^{-61}\right)$, spatial homogeneity on super horizon scales with inhomogeneity of the order 10^{-5} might appear as peculiar initial conditions that require further investigation. Both, the flatness and horizon problem appear under the assumption that the equation of state parameter of the cosmological fluid obeyed $w > -1/3$ at all times. Any form of matter with $w < -1/3$ may thus be an interesting candidate for the solution of these problems and in fact this equation of state can easily be realized by scalar fields.

2. The paradigm of cosmological inflation

Inflation is a cosmological epoch that is assumed to have occurred before the temperature was $T_{reh} \lesssim 10^{16}\text{GeV}$, cf. Bassett et al. (2006), in the the standard Big Bang scenario. A first version of the inflationary theory was presented by Starobinsky (1980) but it became widely known through the work of Guth (1981), which is now called 'Old Inflation'. The idea was further developed by Linde (1982); Albrecht and Steinhardt (1982) and in particular Linde (1983). Linde (2008) provides a recent overview of the subject, while a detailed presentation is given by Baumann (2009).

Inflation now has become the prevailing paradigm for the explanation of spatial flatness, large scale homogeneity and the origin of structure. Generally, inflation is defined as an era in which the comoving hubble radius decreases with time, which is equivalent to acceleration of the scale factor,

$$\text{Inflation} \iff \partial_t (aH)^{-1} < 0 \iff \partial_t^2 a > 0\,.$$

This definition is independent of the underlying theory of Gravity. In General relativity, within the FLRW models, these conditions can be translated into a requirement on the equation of state of the cosmological fluid via Raychaudhuri's equation (1.1),

$$\text{Inflation} \iff \rho + 3p < 0\,. \qquad (2.1)$$

Throughout this chapter we assume $\Lambda = 0$. In the forthcoming sections we will review a standard approach to cosmological inflation.

2.1. Scalar field inflation

The prevailing theories on the cause of inflation rely on scalar fields. In this section we will give a short introduction to the standard approach via a single, minimally coupled, scalar field dominated by its potential energy. Thus the action of the scalar field is

$$S = \int d^4x \sqrt{-g}\mathcal{L}_\phi = \int d^4x \sqrt{-g}\left(-\frac{1}{2}g^{\mu\nu}\partial_\mu\phi\partial_\nu\phi - V(\phi)\right),$$

for which the Euler-Lagrange equations yield

$$\partial_\mu \frac{\partial \mathcal{L}_\phi}{\partial(\partial_\mu \phi)} - \frac{\partial \mathcal{L}_\phi}{\partial \phi} = 0 \implies -\Box\phi + V' = 0,\quad \Box := \nabla_\mu \nabla^\mu\,. \qquad (2.2)$$

From the definition of the stress-energy tensor

$$T_{\mu\nu} := \frac{-2}{\sqrt{-g}}\frac{\partial}{\partial g^{\mu\nu}}\left(\sqrt{-g}\mathcal{L}_{mat}\right),$$

2. The paradigm of cosmological inflation

one obtains
$$T_{\mu\nu} = \partial_\mu\phi\partial_\nu\phi - \left(\frac{1}{2}g^{\alpha\beta}\partial_\alpha\phi\partial_\beta\phi + V\right)g_{\mu\nu}.$$

Einstein's field equations can be derived from the action principle by adding the Einstein-Hilbert Lagrangian $\mathcal{L}_{EH} = \mathcal{R}/(16\pi)$. A comparison with the stress energy tensor of a perfect fluid (1.1) shows that one can relate the scalar field to a perfect fluid by the identification

$$\rho = -\frac{1}{2}g^{\mu\nu}\partial_\mu\phi\partial_\nu\phi + V, \quad p = -\frac{1}{2}g^{\mu\nu}\partial_\mu\phi\partial_\nu\phi - V, \quad u_\mu = \frac{\partial_\mu\phi}{\sqrt{-g^{\mu\nu}\partial_\mu\phi\partial_\nu\phi}},$$

cf. Madsen (1988). Furthermore, given that there is a three-dimensional spatial slice on which the scalar field is constant there exists a foliation of spacetime with RW geometry. In this frame the equation of motion (2.2) reduces to

$$\partial_t^2\phi + 3H\partial_t\phi + V' = 0, \tag{2.3}$$

and the Friedmann equation becomes

$$H^2 + \frac{k}{a^2} = \frac{8\pi}{3}\left(\frac{1}{2}(\partial_t\phi)^2 + V\right). \tag{2.4}$$

The energy density, pressure and velocity associated with the scalar field are

$$\rho = \frac{1}{2}(\partial_t\phi)^2 + V, \quad p = \frac{1}{2}(\partial_t\phi)^2 - V, \quad u^\mu = \delta^\mu_t.$$

Looking at equation (2.1) it is readily seen that inflation occurs when $(\partial_t\phi)^2 < V(\phi)$. If the potential energy is sufficiently large the friction term will dominate equation (2.3), that is $\partial_t^2\phi \ll H\partial_t\phi$, such that the field will change only little on the expansion timescale H^{-1}. Consequently, the energy density of the scalar field is nearly constant which effectively acts as a cosmological constant and leads to quasi exponential growth of the scale factor. This inflationary phase continues for as long as potential energy dominates. It ends with oscillations of the field around the minimum of the potential and its eventual decay gives way to the reheating era that is associated with the begin of the standard Big Bang scenario.

The latter two inequalities are known as the slow-roll regime of a given potential. In fact, in practice it is useful to define the slow-roll parameters

$$\epsilon := \frac{1}{16\pi}\left(\frac{V'}{V}\right)^2 \quad \text{and} \quad \eta := \frac{1}{8\pi}\frac{V''}{V},$$

in order to quantify whether a given potential allows for an extended inflationary phase. Given that these parameters are smaller than unity and the energy density is sufficiently large (but smaller than Planck density) the model provides an inflationary solution.

We can further restrict our qualitative discussion of the solutions to Euclidean spatial sections because curvature quickly becomes negligible due to the quasi exponential expansion of physical scales. In this case, equations (2.3) and (2.4) can be written as a single, nonlinear second-order equation for $\phi(t)$. Moreover, since the resulting equation does not contain t explicitly, it can be reduced to a first-order equation for $v(\phi) := \partial_t \phi$,

$$vv' + v\sqrt{24\pi \left(v^2/2 + V\right)} + V' = 0\,.$$

Solutions of this equation approach an attractor solution exponentially quickly for a large class of potentials $V(\phi)$, cf. Helmer and Winitzki (2006). Thus, regardless of the initial starting point $(\phi, \partial_t \phi)$, the solution essentially evolves along the attractor curve after a short time and this property coined the phrase that inflation 'forgets' its initial conditions.

These properties make an inflationary phase appear ubiquitous in the theory of cosmological scalar fields. However, as we discuss in more detail later on, homogeneity of the scalar field is a crucial assumption and inhomogeneities might prevent the onset of inflation.

Moreover, it should be mentioned that instead of specifying a potential $V(\phi)$ and deriving the corresponding inflationary evolution one could also proceed the other way around. As was worked out by Ellis and Madsen (1991) there exists an algorithm that allows to exactly construct a potential $V(\phi)$ that provides any desired evolution history of the scale factor $a(t)$.

2.1.1. Explanation for flatness, large scale homogeneity and tiny inhomogeneities

Notice that during inflation the equation of state parameter associated with the scalar field obeys $-1 \leq w < -1/3$. Consequently, it follows from equation (1.5) that $\Omega = 1$ becomes an attractor which means that spatial curvature is diminished as inflation proceeds. Moreover, when $w < -1/3$ the comoving particle horizon decreases. This means that the whole surface of last scattering is in causal contact if physical scales expanded by a factor of $\simeq 60$ efolds during the inflationary phase.

Furthermore, density perturbations arise through fluctuations in the scalar field which lead to a local delay of the end of inflation, $\delta t = \delta \phi / \partial_t \phi$. The typical amplitude of the fluctuation during a time interval $t = H^{-1}$ is $|\delta \phi| \simeq H/(2\pi)$, cf. Vilenkin and Ford (1982); Linde (1982). This allows for an estimate of the amplitude of the relative density contrast by

$$\delta_H := \frac{\delta \rho}{\rho} \simeq \frac{\delta t}{t} \simeq \frac{H^2}{2\pi \partial_t \phi}\,.$$

This result was first derived by Mukhanov and Chibisov (1981) and subsequently by Hawking (1982); Starobinsky (1982); Guth and Pi (1982); Bardeen et al. (1983).

2.2. Debate on initial conditions

In the last section it was pointed out that the inflationary process will inevitably occur once the potential energy dominates over the kinetic term. This observation was based

2. The paradigm of cosmological inflation

on the assumption that spacetime obeys RW geometry and therefore raises the question whether inflation can occur in anisotropic or inhomogeneous spacetimes. Since in scalar field inflation the potential energy effectively acts as a cosmological constant the cosmic 'no hair' conjecture has frequently been employed to shed some light on the problem. Roughly speaking, this conjecture states that spacetimes with a positive cosmological constant and expanding spatial sections approach a de Sitter solution asymptotically. This is of course not a precise definition but several rigorous theorems have been proven under this general theme. One of these theorems concerns the evolution of the Bianchi spacetimes which are homogeneous but anisotropic universes was presented by Wald (1983): "All initially expanding Bianchi cosmologies (except Bianchi type IX), containing a positive cosmological constant and matter fields satisfying the dominant and strong energy conditions evolve toward the de Sitter solution on an exponentially rapid time scale." Considering inhomogeneous universes Barrow and Stein-Schabes (1984) found a solution with cosmological constant and pressureless matter that approaches a de Sitter solution locally, i.e. inside the horizon of each observer. Subsequent studies, as summarized by Goldwirth and Piran (1992), have indicated that once the cosmological constant begins to dominate and inflation has started in an anisotropic or inhomogeneous universe, the deviation from isotropy or homogeneity will decay. Moreover, this statement is also valid when scalar field matter is considered, though the respective stress-energy tensor does not obey the strong energy condition during inflation (by definition).

However, these results apply only once inflation is already underway so that they are not sufficient to answer the question of whether inflation begins in the first place under general anisotropic or inhomogeneous initial conditions.

For the case of anisotropic initial conditions Rothman and Ellis (1986) show that, within certain caveats, anisotropy generally does not prevent but possibly enhances inflation. Raychaudhuri and Modak (1988) however conclude that: "one cannot reach any definitive conclusion - all that one can say is that in a universe in which different regions exhibit all possible behaviour, some regions would indeed undergo inflation." For inhomogeneous initial conditions the question was addressed numerically by Goldwirth and Piran (1990). They find that the scalar field must have a value suitable for inflation across several horizon lengths in order for inflation to begin. The result was confirmed with the analytical approach by Calzetta and Sakellariadou (1992). The most general study by Vachaspati and Trodden (2000) proves and generalizes these findings: "Inflationary models based on the classical Einstein equations, the null energy conditions, and trivial topology, require initial homogeneity on super-Hubble scales." Accordingly, 'local' inflation is not possible under these assumptions.

3. Eternal inflation, vacuum decay and bubble collisions

3.1. Eternal inflation

The essence of eternal inflation is that inflation may never end in the whole spacetime. The general idea was developed already in the early works on scalar field inflation when it was realized that the field fluctuations have a 'selfreproducing' regime. The evolution of the scalar field depends on how the fields random fluctuations compare to its classical background trajectory. When the fluctuations are dominated by the deterministic change, $\delta\phi \ll \partial_t\phi H^{-1}$, the evolution proceeds essentially unaltered. However, when the classical change is much smaller than the fluctuations, $\delta\phi \gg \partial_t\phi H^{-1}$, the fields evolution resembles a random walk. Consequently, if the field is approximately homogeneous across several horizon sizes, cf. the discussion in the preceeding section, this region will continue to inflate and potentially contains further fluctuation dominated regions, thereby 'reproducing itself'. In this picture inflation ends only locally, but proceeds forever globally. Strictly speaking, the term eternal inflation is associated with inflationary models that continue indefinitely into the future but not to the past as it has been argued by Borde et al. (2003) that inflation cannot be 'past-eternal'.

The phenomenon of eternal inflation also occurs in scalar field inflation if the potential has multiple local minima. In fact, the very first model of inflation by Guth (1981) involved a scalar field in a so-called false vacuum, that is a local, but not global minimum of the potential. A quantum mechanical consideration of the configuration, cf. Coleman (1977), shows that there is a non-vanishing probability for the field to tunnel to another minimum with lower potential energy, thereby spawning regions (bubbles) of lower potential energy, with energy confined to the domain wall. These bubbles appear randomly at various places and times, with a fixed rate per unit four-volume. Accordingly, in Guth's original scenario inflation proceeds for as long as the field is stuck in a false vacuum. Eventually, the collision of bubbles releases the energy of the domain walls so that the universe becomes hot and passes over to the standard Hot Big Bang scenario. However, this model of inflation is somewhat misleading. If the tunneling probability is large, then bubbles nucleate near to each other and collide immediately so that the inflationary phase is too short to solve any problems. If, on the other hand, the probability of bubble formation is low, the bubbles do not collide and each of the bubbles resembles a separate empty open universe. Both options are unacceptable, which has lead to the conclusion that this scenario, termed 'new inflation', cannot explain the inflationary process. Consequently, the model was abandoned with the advent of the 'chaotic inflation' scenario by Linde (1983).

3. Eternal inflation, vacuum decay and bubble collisions

Spurred by astronomical evidence for $\Omega < 1$ in the mid-1990s the idea of false vacuum inflation and vacuum decay regained some interest. This originated in the fact that the spacetime that represents the interior of the bubble can be foliated into homogeneous spacelike slices on which the scalar field is constant and that these spatial sections have a hyperbolic geometry. However, interest in these so-called open inflation scenarios, e.g. Gott (1982); Bucher et al. (1995); Yamamoto et al. (1995); Linde (1999), has weakened due to the fact that recent observations imply $\Omega \simeq 1$, though the possibility that space has a hyperbolic geometry is not yet excluded.

Eternal inflation and false vacuum tunneling have then once more returned to the fore with the emergence of the 'landscape' idea in string theory, cf. Bousso and Polchinski (2000); Susskind (2003); Kachru et al. (2003). A major driving force is the so-called 'cosmological constant problem', cf. Weinberg (1989); Carroll (2001); Padmanabhan (2003); Peebles and Ratra (2003); Ellis (2003); Bousso (2008), that the value which is ascribed to the cosmological constant is tiny, $\Lambda \approx 10^{-120}$. Attempts to calculate the energy density of the vacuum in quantum field theory 'naturally' suggest $\Lambda \approx 1$. String theorists argue that the problem of the smallness of the cosmological constant can be addressed within the landscape paradigm. The landscape involves a very large number ($\gtrsim 10^{500}$) of metastable vacua corresponding to the minima of the effective potential in field space which differ in the geometry of spacetime compactification. The value of the potential at each minimum is the effective value of the cosmological constant in the corresponding vacuum. Consequently, arbitrarily many bubbles with different values of their interior vacuum energy density can be created by the tunneling process. Vacua with $\Lambda \leq 0$ do not allow any further tunneling, whereas vacua with $\Lambda > 0$ keep up the decay chain as they give rise to many nested successor bubbles. It is thus an appealing idea that we might just happen to live in a region of space which underwent many such transitions in the past leading to a net decrease in vacuum energy density so that it is already 'close' to the final state $\Lambda = 0$.

Unfortunately, string theory does not specify a single 'preferred' vacuum state so that one is forced to determine the probability distribution of vacua as found by a randomly chosen observer, an intricate task that is known as the 'measure problem' in eternal inflation. The existing proposals for a probability measure fall in two classes, designated as 'volume-based' and 'worldline-based'. The difference between these classes of measures is in the approach taken to construct the ensemble of observers. In the volume-based approach, the ensemble contains every observer appearing in spacetime. In the worldline-based approach the ensemble consists of observers appearing near a single, randomly selected timelike geodesic.

More details and a thorough presentation of eternal inflation can be found in the textbook by Winitzki (2009). Moreover, the collection of articles by Carr (2008) provides a comprehensive overview on the debate on the uniqueness of the universe. In the next section we will briefly indicate the field theoretic tunneling mechanism in Minkowski spacetime.

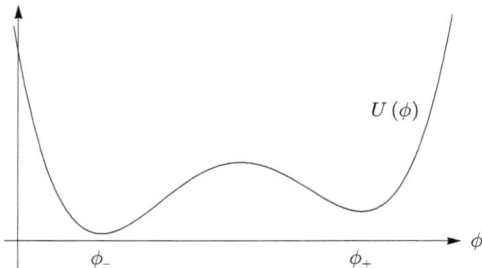

Figure 3.1.: Illustration of a scalar field potential $U(\phi$ with two minima that allows for quantum tunneling. The state $\phi = \phi_+$ is called a false vacuum because it is not a stable minimum due to decay to $\phi = \phi_-$ by quantum tunneling.

3.2. Decay of a metastable vacuum

A theory of vacuum decay with the neglection of gravitational effects based on instanton methods was presented by Coleman (1977). In this section we will review the basic arguments that allow to calculate a nucleation rate for vacuum bubbles in Minkowski spacetime.

Consider a theory of a single scalar field with action given by equation (2.1). When the potential $U(\phi)$ has two minima, say at $\phi = \phi_-$ and $\phi = \phi_+$, with $\epsilon := U(\phi_+) - U(\phi_-)$, the state $\phi = \phi_+$ is called a false vacuum because it is not a stable state due to decay by quantum tunneling, cf. Figure 3.1. The tunneling rate $\Gamma = A \exp\left(\mathcal{S}_E(\phi_+) - \mathcal{S}_E(\phi)\right)$, is given by the Euclidean action $\mathcal{S}_E(\phi)$ of the field configuration ϕ. The prefactor A is difficult to calculate and is usually estimated to be of the order of unity. It can be shown that a solution to the corresponding Euler-Lagrange equations must be $O(4)$ symmetric, so that ϕ is a function only of $\bar{r}^2 := -t^2 + \mathbf{x}^2$, the Euclidean distance from an appropriately chosen center of coordinates. Accordingly, the simplified Euclidean action takes the form

$$\mathcal{S}_E = 2\pi^2 \int d\bar{r}\, \bar{r}^3 \left(\frac{1}{2}(\partial_{\bar{r}}\phi)^2 + U\right),$$

with the equation of motion

$$\partial_{\bar{r}}^2 \phi + \frac{3}{\bar{r}}\partial_{\bar{r}}\phi - U' = 0\,.$$

subject to the boundary conditions

$$\lim_{\bar{r}\to 0} \phi(\bar{r}) = \phi_-, \quad \lim_{\bar{r}\to\infty} \phi(\bar{r}) = \phi_+, \quad \partial_{\bar{r}}\phi\big|_{\bar{r}=0} = 0\,.$$

3. Eternal inflation, vacuum decay and bubble collisions

Formally, a first integral of the equation of motion is

$$\frac{1}{2}(\partial_{\bar{r}}\phi)^2 - (U(\phi) - U(\phi_+)) = \int_{\bar{r}}^{\infty} d\bar{r}' \frac{3}{\bar{r}'}(\partial_{\bar{r}'}\phi)^2 \quad \text{and} \quad \epsilon = \int_0^{\infty} d\bar{r}' \frac{3}{\bar{r}'}(\partial_{\bar{r}'}\phi)^2 .$$

To proceed, one makes a thin wall approximation, which means that one requires that the field is nearly constant and equal to ϕ_- within the bubble of radius $\bar{r}_0$, makes a sharp step within a thin layer (the bubble wall), and tends to ϕ_+ outside the bubble. This means $(\partial_{\bar{r}}\phi)^2/2 \simeq U(\phi) - U(\phi_+)$ within the bubble wall and zero otherwise. Moreover, it implies $\epsilon \simeq 3\sigma/\bar{r}_0$ where

$$\sigma := \int_0^{\infty} d\bar{r} (\partial_{\bar{r}}\phi)^2 \simeq \int_{\phi_-}^{\phi_+} d\phi \sqrt{2(U(\phi) - U(\phi_+))} ,$$

defines the surface energy density of the domain wall. Consequently, the action becomes

$$\mathcal{S}_E(\phi) - \mathcal{S}_E(\phi_+) \simeq -\frac{\pi^2}{2}\epsilon \bar{r}_0^4 + 2\pi^2 \sigma \bar{r}_0^3 \simeq \frac{27\pi^2 \sigma^4}{2\epsilon^3} ,$$

which determines the tunneling rate in the semiclassical approximation. The interpretation is that the bubble 'materializes' in Minkowski spacetime, with its wall expanding into the surrounding space and thereby converting the old vacuum to the new. The bubble wall quickly accelerates and reaches essentially the speed of light so that it traces out the hyperboloid $\bar{r}^2 = -t^2 + \mathbf{x}^2$.

The interior of the hyperboloid may be foliated by 'infinite', (though cf. Ellis and Stoeger (2009)), uniform, hyperbolic spatial sections, which constitute a hyperbolic RW geometry. Given that the scalar field does not tunnel directly to the minimum of the potential but slightly uphill, inflation, as it is explained in the last chapter, may occur within the bubble in which case it is called open inflation, cf. Bucher et al. (1995); Yamamoto et al. (1995); Linde (1999) and the illustration in Figure 3.2.

3.3. Classical bubble evolution

This approach was extended for the inclusion of gravity by Coleman and de Luccia (1980). The subsequent classical evolution of bubbles in a vacuum has been studied in detail by Blau et al. (1987); Berezin et al. (1987); Aurilia et al. (1989); Aguirre and Johnson (2005). Their analysis of the bubble evolution is based on the Israel junction method, which is explained in Appendix D. The method allows to join different solutions of Einstein's equations such that Einstein's equations are also valid in the resulting spacetime. It is thus possible to make a junction of a bubble with vacuum energy density Λ^- and surface energy density σ and a vacuum background with vacuum energy density Λ^+. The involved spacetimes are de Sitter and Schwarzschild – de Sitter respectively. For convenience we may identify

$$\Lambda^+ := U(\phi_+), \quad \Lambda^- := U(\phi_-), \quad k := 4\pi\sigma .$$

3.3. Classical bubble evolution

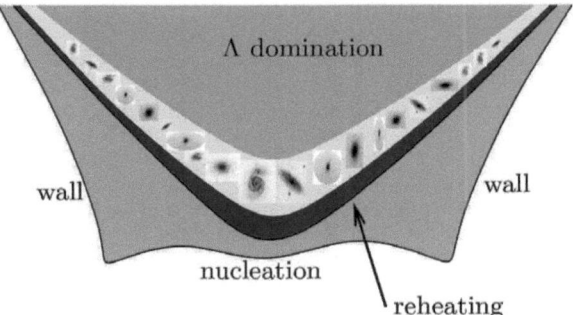

Figure 3.2.: Illustration of how the interior of a bubble, nucleated by quantum tunneling of the scalar field, may give rise to inflation and the subsequent cosmological evolution. Figure from Winitzki (2008).

It follows from the junction conditions that the Schwarzschild mass M in the exterior spacetime is related to the surface energy density σ, the areal radius R and the proper velocity $\dot{R}$. It turns out to be useful to introduce dimensionless variables z and T for the areal radius R and the proper time τ on the bubble via

$$z := \sqrt[3]{\frac{L^2}{2M}} R, \quad T := \frac{L^2}{2k}\tau, \quad L^2 := \frac{1}{3}\sqrt{\left|(\Lambda^+ + \Lambda^- + 3k^2)^2 - 4\Lambda^+\Lambda^-\right|}.$$

The evolution equation of the areal radius of the bubble wall can be cast into the form of the equation of motion of a point particle of unit mass in a one-dimensional potential

$$\left(\frac{dz}{dT}\right)^2 + V(z) = Q, \quad V := -\left(z^2 + \frac{2Y}{z} + \frac{1}{z^4}\right),$$
$$Q := -\frac{4k^2}{(2M)^{2/3} L^{8/3}}, \quad Y := \frac{\Lambda^+ - \Lambda^- + 3k^2}{3L^2}. \tag{3.1}$$

The solutions depend solely on the parameters Λ^+, Λ^-, M and were studied and catalogued by Blau et al. (1987); Aguirre and Johnson (2006).

The underlying assumption of this picture is that the bubble nucleates in a vacuum background. The calculation of the probability of bubble nucleation in matter dominated, or inhomogeneous backgrounds is a complicate task and has rarely been addressed in the literature, cf. Abbott et al. (1987). However, given that bubble nucleation is possible in such backgrounds it would be interesting to figure out how the presence of matter, or inhomogeneity, affects the subsequent evolution of the bubble, a question that can well be addressed within the aforementioned junction approach. The first part of the present thesis is therefore devoted to the study of the evolution of vacuum bubbles in non-vacuum backgrounds.

3.4. Observability of bubble collisions

Another aspect in the evolution of vacuum bubbles is the possibility of bubble collisions some implications of which were studied already by Hawking (1982); Guth and Weinberg (1983). Shortly thereafter Gott and Statler (1984) proposed to constrain inflationary models by requiring that they do not allow for a bubble collision in our past lightcone as it was then generally assumed that those were incompatible with cosmological observations. This preconception was questioned by Garriga et al. (2007) who showed that the expected number of bubble collisions in the past lightcone of an observer in eternal inflation depends on the position of the observer within the bubble, thereby defining a 'center' of the bubble and a preferred frame for the overall bubble distribution. This work inspired a series of papers by Aguirre et al. (2007); Freivogel et al. (2007); Chang et al. (2008); Aguirre and Johnson (2008); Aguirre et al. (2009); Freivogel et al. (2009) who were the first to assume that bubble collisions might exist in our past lightcone. An important conclusion of these works is that, if compatible with observations, there is no reason to expect that a collision region with angular scale 2π is not causally accessible. However, the signatures of the collision are stretched during the inflationary epoch so that we must assume that inflation does not last much longer than required to satisfy the observational bound on Ω_k. Though this might be considered as tuning there is some indication that the landscape scenario even favors a small number of inflationary e-folds, cf. Freivogel et al. (2006).

Furthermore, under a variety of approximations, Chang et al. (2009) have indicated how a bubble collision leads to a shift in the reheating surface, i.e. the spatial slice where slow roll of the scalar field comes to an end, and how this shift translates to a perturbation in the surface of last scattering and consequently the CMB. They point out that a bubble collision can in principle be seen as hot/cold spots in the CMB. The spots must obey azimuthal symmetry as a consequence of the $SO(2,1)$ symmetry of the spacetime describing the collision of two vacuum bubbles, cf. Garriga et al. (2007). This opens up the remarkable possibility that the dynamics of false vacuum eternal inflation are accessible to observational cosmology.

In general, a modulation of the CMB fluctuations may be decomposed as

$$\frac{\Delta T(\hat{\mathbf{n}})}{T_0} = (1 + f(\hat{\mathbf{n}}))(1 + \delta(\hat{\mathbf{n}})) - 1,$$

cf. Gordon et al. (2005). Here $f(\hat{\mathbf{n}})$ is the signal induced by the collision and $\delta(\hat{\mathbf{n}})$ represents the effect of the fluctuations of the scalar field in direction $\hat{\mathbf{n}}$. In particular, Chang et al. (2009) have derived a signal of the form

$$f(\hat{\mathbf{n}}) = (c_0 + c_1 \cos\vartheta) \cdot \Theta(\vartheta_D - \vartheta). \tag{3.2}$$

A similar kind of signal is at the heart of the second part of the thesis where we ask whether such spots can possibly be identified with the help of Minkowski functional statistics of the CMB temperature map.

A status report on the observability of cosmic bubble collisions that contains many further references is to be found in Aguirre and Johnson (2009).

Part I.
Vacuum Bubbles on Dynamical Backgrounds

4. Introduction to Part I

Recently, papers have accumulated which suggest that tunneling rates in the 'landscape' of string theory can be high, so that lifetimes of metastable vacua can be smaller than a Hubble time. In particular Tye (2006) and Tye and Wohns (2009) have argued that tunneling rates can be enhanced due to resonant tunneling. Moreover, in the context of Dirac-Born-Infeld inflation, Brown et al. (2007) have found that the decay rate can become orders of magnitude larger than in the Coleman-De Luccia prediction. Consequently, one may question the effectivity of the cosmic no-hair mechanism in this case and ask whether bubbles may form in backgrounds which are not vacuum dominated but contain a dynamically relevant amount of matter. In fact, inhomogeneous initial states may even further enhance tunneling rates, cf. Saffin et al. (2008). This suggests to consider tunneling and evolution of vacuum bubbles in non-vacuum and inhomogeneous backgrounds which is the subject of this part of the thesis, cf. Figure 4.1. The machinery to study bubble evolution in inhomogeneous backgrounds was already set up by Berezin et al. (1987) and in particular by Fischler et al. (2008). Here we will elaborate on this approach and study the effects of surrounding matter and inhomogeneity.

We consider a spherical bubble with vacuum energy density Λ^- moving into a region of higher vacuum energy density Λ^+ and assume that the region that separates the different vacua is small compared to the physical radius of the bubble, such that it can be treated as a domain wall. This corresponds to the so-called thin shell approximation. Further, we assume that the interior of the bubble is represented by de Sitter spacetime, while the exterior spacetimes should allow for the presence of matter. Thus, as exterior spacetimes, we will consider solutions of Friedmann-Robertson-Walker type, or, for radially inhomogeneous dust or curvature profiles, of Lemaître-Tolman type. It is the inclusion of matter and curvature in the background that is the new ingredient that is studied in the present work. As will be shown, already the presence of matter has important consequences for the evolution of the domain wall. Moreover, it turns out that curvature inhomogeneities as well as rapid phase transitions in the background affect the motion of the domain wall.

This part of the thesis contains an introduction to the basic properties of the Lemaître-Tolman spacetime in chapter 5. This spacetime is matched to a de Sitter solution along a common spherically symmetric boundary, the domain wall. We require that this junction is a valid solution to Einstein's field equations which means that the Israel junction conditions are to be fullfilled. These will be discussed for the Lemaître-Tolman and Friedmann-Robertson-Walker spacetimes in a unified manner in chapter 6. Chapter 7 presents an extensive analysis of the evolution of the domain wall on a variety of different exterior spacetimes. We start with a brief review of the evolution on a vacuum background and then proceed with the study of the domain wall's motion on homogeneous

4. Introduction to Part I

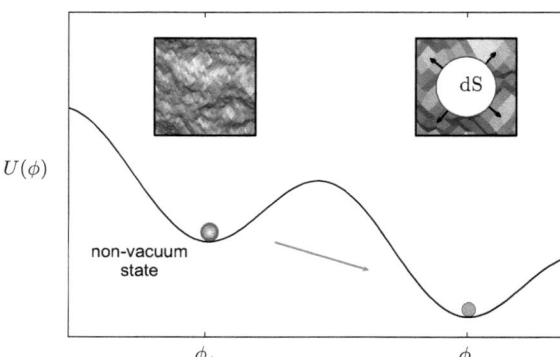

Figure 4.1.: Illustration of the tunneling process in a not yet vacuum dominated region of space. The scalar field rests at $\phi = \phi_+$ and thus effectively acts as a cosmological constant. We assume that, as a vacuum bubble nucleates, the background is dominated by another form of matters which alters the subsequent evolution of the bubble in comparison to the vacuum case.

dust backgrounds, flat backgrounds with a radially increasing dust profile, backgrounds with radial inhomogeneities in curvature and dust density and in a background with homogeneous matter that undergoes a rapid phase transition. In the last chapter of this part we summarize and draw conclusions about the influence of a dynamical background on the evolution of a domain wall.

5. The Lemaître-Tolman spacetime

The Lemaître-Tolman spacetime, (Lemaître (1933); Tolman (1934)), is a spherically symmetric spacetime with a dust source. In the absence of pressure the acceleration of the dust particles vanishes, i. e. $u^\mu \nabla_\mu u^\nu = 0$, which means that the dust particles move on timelike geodesics. This allows to rescale the time coordinate of the comoving synchronous coordinate system such that t is the proper time of the dust particles. Hence, by virtue of Einstein's equations the line element of a spherically symmetric spacetime can be simplified to

$$ds^2 = -dt^2 + \frac{(\partial_r R(t,r))^2}{1+2E(r)} dr^2 + R^2(t,r) d\Omega^2. \tag{5.1}$$

Details of the derivation are provided in Appendix B. Here E is an arbitrary function constrained to $E \geq -1/2$ to avoid a change in the signature of the metric. From Einstein's equations the evolution of R is determined by

$$(\partial_t R)^2 = 2E + \frac{2M}{R} + \frac{\Lambda}{3} R^2, \tag{5.2}$$

where $M(r)$ is another arbitrary function. The meaning of M can be read off from the other non-trivial Einstein equation which determines the evolution of the energy density ρ,

$$8\pi\rho = \frac{2\partial_r M}{R^2 \partial_r R}, \tag{5.3}$$

Thus $M(r)$ can be interpreted as the active gravitational mass contained in the shell of constant coordinate radius r. Note that, unless $E = 0$, this is not equal to the corresponding volume integral of the energy density. In regions where $E < 0$ the active gravitational mass is less than the sum of its constituents while it is greater where $E > 0$. This vivid illustration of gravitational binding energy was at first pointed out in Bondi (1947).

The formal solution of equation (5.2) is

$$\int_0^R \frac{d\tilde{R}}{\sqrt{2E + \frac{2M}{\tilde{R}} + \frac{\Lambda}{3}\tilde{R}^2}} = t - t_B. \tag{5.4}$$

Here $t_B(r)$, the bang time function, is a third arbitrary function that defines the position dependent time of the big bang. Thus, the Lemaître-Tolman spacetime has three free functions E, M, t_B that have to be specified in order to fix the model completely. However, the form of equation (5.1) is invariant under changes of the radial coordinate $\tilde{r} = f(r)$, so there is also a degree of freedom in picking coordinates.

5. The Lemaître-Tolman spacetime

In general, the integral in equation (5.4) involves elliptic functions but a parametric solution can be given when $\Lambda = 0$. The solution depends on the sign of E and is classified accordingly.

- Elliptic, $E < 0$:

$$R = \frac{M}{-2E}(1 - \cos\eta) , \qquad (5.5a)$$

$$t - t_B = \frac{M}{(-2E)^{3/2}}(\eta - \sin\eta) . \qquad (5.5b)$$

The crunch time is $t_C := t_B + 2\pi M/(-2E)^{2/3}$.

- Parabolic, $E = 0$:

$$R = \left[\frac{9}{2}M(t - t_B)^2\right]^{1/3} . \qquad (5.6)$$

- Hyperbolic, $E > 0$:

$$R = \frac{M}{2E}(\cosh\eta - 1) , \qquad (5.7a)$$

$$t - t_B = \frac{M}{(2E)^{3/2}}(\sinh\eta - \eta) . \qquad (5.7b)$$

The dimensionless quantity η is defined by $\eta(t, r) := \sqrt{|2E(r)|}\int_{t_B}^{t} d\tilde{t}/R(\tilde{t}, r)$. When E changes sign these equations may still be applicable and the local time evolution is governed by the local sign of E. For example, it is perfectly possible to have a negatively curved, expanding space in between two positively curved regions that undergo gravitational collapse, see Figure 5.1. This and other examples of curious properties of the Lemaître-Tolman spacetime can be found in Plebanski and Krasinski (2006). Furthermore, in the limit of small R, the solutions with $E \neq 0$ reduce to the parabolic case (5.6).

In the subcase $M = $ constant the dust source vanishes. Then, depending on the presence of a cosmological constant, the Lemaître-Tolman spacetime corresponds to the Kottler (1918), or Szekeres-Kruskal-Schwarzschild spacetime, (Szekeres (1960); Kruskal (1960); Schwarzschild (1916)), in geodesic coordinates. Here, as was shown by Hellaby (1996), the choice of the functions E and t_B determines which part of the manifold is covered.

With a constant bang time function and $E/M^{2/3} = $ constant the Lemaître-Tolman spacetime reduces to the Friedmann-Robertson-Walker spacetime Friedmann (1924).

Shell-crossing singularities

The dust density defined by equation (5.3) becomes infinite where $R = 0 \neq \partial_r M$ and where $\partial_r R = 0 \neq \partial_r M$. The former corresponds to the big bang singularity at $t = t_B$,

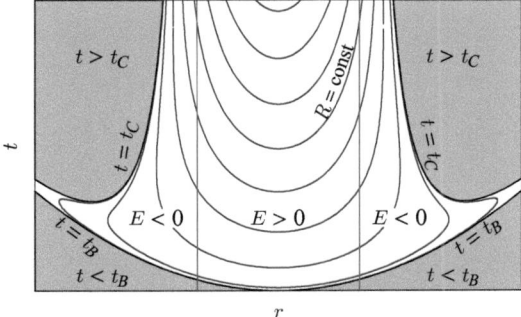

Figure 5.1.: Diagram of a Lemaître-Tolman spacetime with a sign changing curvature and $\Lambda = 0$. Along the blue lines we have $R = $ constant. At any instant $t > t_B$ there are two centers of symmetry. Regions to the left and to the right of the vertical red lines have $E < 0$ and experience gravitational collapse, while we have $E > 0$ in between where spatial sections expand at all times. The contour $R = 0$ represents the boundary of the spacetime, part of it are the big bang and big crunch surfaces.

while the latter represents the occurrence of a so called shell-crossing singularity. At a shell-crossing singularity the physical distance between two adjacent dust shells vanishes which enforces the divergence of the matter density. Shell crossings are considered as artifacts in the Lemaître-Tolman solution because it is expected that in real astrophysical situations, as the matter becomes more dense, pressure gradients arise and prevent the shells from collision.

6. Spacetime junction

6.1. General considerations

In this section we will formulate the spacetime junction conditions, as introduced in Appendix D, that arise in the course of matching the Lemaître-Tolman (LT), or Friedmann-Lemaître-Robertson-Walker (FLRW) to a de Sitter (dS) spacetime across a timelike, spherically symmetric hypersurface Σ. This hypersurface shall represent the domain wall that separates the regions of different vacua and matter content. The line element of the hypersurface is

$$ds_\Sigma^2 = -d\tau^2 + \bar{R}^2(\tau) d\Omega^2 \,. \tag{6.1}$$

Here τ is the proper time on the shell and $\bar{R}(\tau)$ represents the time evolution of the areal radius. We refrain from introducing Gaussian normal coordinates but instead identify each $\tau = $ constant slice of Σ with a $(t = \text{constant}, r = \text{constant})$ hypersurface in $\mathcal{M}^+$ and $\mathcal{M}^-$. Hence, with identification of the angular variables, the worldsheet of the bubble in $\mathcal{M}^\pm$ can be parameterized as $(\bar{t}_\pm, \bar{r}_\pm)$. Henceforth an overbar shall represent the evaluation of a quantity on Σ and for clarity the index $\pm$ will be omitted when possible. The basis vectors on Σ are

$$e_\tau^\mu = \dot{\bar{t}} \delta_t^\mu + \dot{\bar{r}} \delta_r^\mu, \quad e_\vartheta^\mu = \delta_\vartheta^\mu \quad \text{and} \quad e_\varphi^\mu = \delta_\varphi^\mu \,,$$

where a dot refers to a partial derivative with respect to τ. Consequently, the first junction (D.1) becomes

$$\dot{\bar{t}}^2 = 1 + \frac{\left(\partial_{\bar{r}} \bar{R}\right)^2}{1 + 2\bar{E}} \dot{\bar{r}}^2 \,, \tag{6.2}$$

and we choose the proper time of the bubble such that $\dot{\bar{t}} > 0$. The normal vector is

$$n_\mu = s \frac{\partial_{\bar{r}} \bar{R}}{\sqrt{1 + 2\bar{E}}} \left(\dot{\bar{t}} \delta_\mu^r - \dot{\bar{r}} \delta_\mu^t \right), \quad \text{with } s = \pm 1 \,. \tag{6.3}$$

It is defined up to a sign only. This determines how $\mathcal{M}^-$ and $\mathcal{M}^+$ are 'sticked together'.

The respective conditions for the FLRW and the dS part follow from the LT expressions by setting $2E(r) = -kr^2$, with $k = $ constant, $t_B = $ constant and $M(r) \propto r^3$.

For the second junction we note that due to spherical symmetry only the two components K_τ^τ and K_ϑ^ϑ of the extrinsic curvature tensor are independent. Therefore the stress-energy tensor on the shell has to be of perfect fluid form and this motivates to define $S_\tau^\tau := -\sigma$ and $S_\vartheta^\vartheta := P$. We will take the most conservative choice in the equation state of the shell by setting

$$\boxed{P = -\sigma \,,}$$

6. Spacetime junction

Thus, the energy density of the domain wall is of vacuum type and there is no additional matter present on the surface. Of course, one might consider other equations of state, for example Fischler et al. (2008) assume a linear equation of state $P = w\sigma$ and explore the corresponding bubble evolution for several values of w.

Having fixed the equation of state of the domain wall the conditions (D.2) reduce to

$$4\pi\sigma = \left[K^\vartheta_\vartheta\right], \tag{6.4a}$$

$$-8\pi\sigma = \left[K^\tau_\tau + K^\vartheta_\vartheta\right]. \tag{6.4b}$$

Similarly, equations (D.4a) and (D.4b) become

$$\dot{\sigma} = \left[T_{\alpha\beta}e^\alpha_\tau n^\beta\right], \tag{6.5}$$

where $T^\pm_{\alpha\beta}$ are the components of the respective stress-energy tensors. Equation (6.5) is an integrability condition for (6.4a) and (6.4b), so only two of these three equations are independent. Therefore, we will only solve the equations (6.4a) and (6.5), with (6.4b) being fulfilled identically. Hence, we rewrite the $\vartheta\vartheta$-component of equation (D.3) with the help of the first junction condition (6.2) and (5.2) to arrive at

$$K_{\vartheta\vartheta} = -s\bar{R}\sqrt{\dot{\bar{R}}^2 + 1 - \frac{2\bar{M}}{\bar{R}} - \frac{\Lambda}{3}\bar{R}^2},$$

Using this in equation (6.4a) gives the equation of motion for $\bar{R}$

$$4\pi\sigma\bar{R} = s_-\sqrt{\dot{\bar{R}}^2 + 1 - \frac{\Lambda^-}{3}\bar{R}^2} - s_+\sqrt{\dot{\bar{R}}^2 + 1 - \frac{2\bar{M}}{\bar{R}} - \frac{\Lambda^+}{3}\bar{R}^2}. \tag{6.6}$$

Solving for the derivative we obtain the more convenient form

$$\dot{\bar{R}}^2 + 2V = -1, \quad \text{with} \tag{6.7}$$

$$2V := -\left[\frac{\Lambda^-}{3} + \left(\frac{\Lambda^+ - \Lambda^-}{24\pi\sigma} + 2\pi\sigma\right)^2\right]\bar{R}^2 - \left(1 + \frac{\Lambda^+ - \Lambda^-}{48\pi^2\sigma^2}\right)\frac{M}{\bar{R}} - \frac{M^2}{16\pi^2\sigma^2\bar{R}^4}.$$

As noted earlier one obtains the Schwarzschild-de Sitter spacetime when $M = $ constant, cf. equation (3.1). In this case all coefficients in the potential are constant and the areal radius of the bubble behaves as a point particle under the potential V. The possible spacetime junctions have been cataloged and discussed in detail in Blau et al. (1987); Aurilia et al. (1989); Aguirre and Johnson (2005).

However, we are going to include matter in the background and therefore M will not be independent of the radial coordinate r. As the bubble moves it will experience different values of M, the potential becomes time dependent. In this way its evolution becomes sensitive to the dynamics of the background.

6.2. Explicit choice of the junction geometry

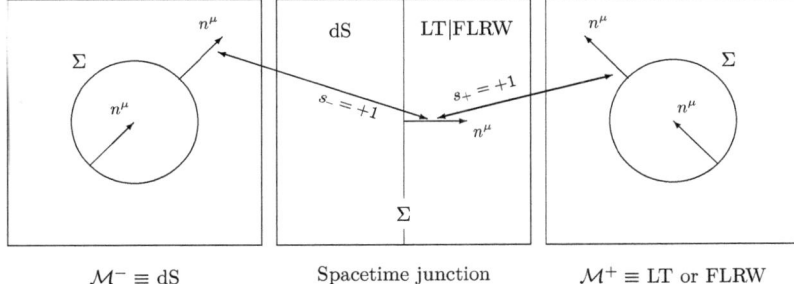

Figure 6.1.: Geometry of the spacetime junction. The boxes show slices of constant time of the junction spacetimes (one dimension suppressed). The circles represent the junction surface Σ. The normal vector is defined to point from $\mathcal{M}^-$ to $\mathcal{M}^+$. The sign s in the definition (6.3) determines how the spacetimes are sticked together. For $s = 1$ the coordinate radius increases in the normal direction, while it decreases when $s = -1$. The arrows indicate the choice in the present work, $s_- = s_+ = 1$.

6.2. Explicit choice of the junction geometry

To proceed we will now pick the junction geometry by fixing the sign of the normal vector. The possible junction geometries depend on the vacuum energy and matter content in the background and on the shell and are determined by equation (6.6). For $\Delta := 2\bar{M}/\bar{R} + (\Lambda^+ - \Lambda^-)\bar{R}^2/3$ these are

$$\Delta < -\left(4\pi\sigma\bar{R}\right)^2 \Leftrightarrow s_+ = -1 \text{ and } s_- = -1, \tag{6.8a}$$

$$-\left(4\pi\sigma\bar{R}\right)^2 < \Delta < +\left(4\pi\sigma\bar{R}\right)^2 \Leftrightarrow s_+ = -1 \text{ and } s_- = +1, \tag{6.8b}$$

$$+\left(4\pi\sigma\bar{R}\right)^2 < \Delta \Leftrightarrow s_+ = +1 \text{ and } s_- = +1. \tag{6.8c}$$

Though all these geometries are possible in principle, we will pick the physically most intuitive configuration, that is

$$\boxed{s_- = s_+ = 1,}$$

which we have attempted to illustrate in Figure 6.1. Observers in $\mathcal{M}^-$ are surrounded by the domain wall and therefore $\mathcal{M}^-$ can be regarded as the interior of the bubble. However, observers in $\mathcal{M}^+$ can in principle encircle the domain wall, so it seems reasonable to regard $\mathcal{M}^+$ as the exterior of the bubble. An immediate consequence of this

6. Spacetime junction

choice and equation (6.8c) is that the surface energy is restricted to

$$\sigma < \frac{1}{4\pi}\sqrt{\frac{2\bar{M}}{\bar{R}^3} + \frac{\Lambda^+ - \Lambda^-}{3}} =: \sigma_{max}. \qquad (6.9)$$

Furthermore we require

$$\boxed{\partial_r R > 0 \quad \text{and} \quad \partial_r M > 0.}$$

The latter allows to choose a coordinate system such that

$$\boxed{M(r) \equiv \frac{4\pi}{3}\varrho r^3,}$$

where ϱ is a constant. For simplicity we also introduce the 'scale factor' $a(t,r) := R(t,r)/r$ which renders the potential in (6.7) to

$$2V = -\left[\frac{\Lambda^-}{3} + \left(\frac{\varrho}{3\bar{a}^3\sigma} + \frac{\Lambda^+ - \Lambda^-}{24\pi\sigma} + 2\pi\sigma\right)^2\right]\bar{R}^2. \qquad (6.10)$$

This equation completes the simplification of equation (6.7) and together with (6.5) provides the basis for the further analysis of the bubble evolution in the exterior coordinates.

6.3. System of equations to be solved

The potential in equation (6.10) will in general be time dependent through the evolution of the background. Therefore we have to establish the background solution before we can go on studying the evolution of a bubble in matter environments. In the following we will briefly summarize the necessary ingredients that determine the dynamics of the background spacetimes.

6.3.1. Specification of the LT spacetime

With the definitions of the preceding section, the equations of motion of the LT spacetime (5.2) and (5.3) become

$$\left(\frac{\partial_t a}{a}\right)^2 = \frac{2E(r)}{a^2 r^2} + \frac{8\pi}{3}\frac{\varrho}{a^3} + \frac{\Lambda^+}{3}, \qquad (6.11a)$$

$$\rho = \frac{\varrho}{a^2(r\partial_r a + a)}. \qquad (6.11b)$$

To solve the background dynamics (6.11a) we have to specify ϱ, Λ^+, a curvature profile $E(r)$ and the 'scale factor' $a_0(r) := a(t_0, r)$. In a non-vacuum background the latter can be substituted by the dust density $\rho_0(r) := \rho(t_0, r)$ via equation (6.11b)

$$a_0 = \left(\frac{3\varrho}{r^3}\int_0^r \frac{\tilde{r}^2 d\tilde{r}}{\rho_0(\tilde{r})}\right)^{1/3}. \qquad (6.12)$$

Hence, the LT background is fixed by the choice of $\{\varrho, \rho_0(r), \Lambda^+, E(r)\}$.

6.3.2. Specification of the FLRW spacetime

For the study of the evolution of the domain wall on a background filled with a perfect fluid that undergoes a rapid phase transition a FLRW solution will be used. It is determined by

$$\left(\frac{\partial_t a}{a}\right)^2 = \frac{8\pi}{3}\rho + \frac{\Lambda^+}{3}, \tag{6.13a}$$

$$\partial_t \rho + \frac{3\partial_t a}{a}(\rho + p) = 0, \tag{6.13b}$$

Here we have to specify the vacuum energy density Λ^+ and the matter density $\rho_0 := \rho(t_0)$ but we will desist from studying any effects of curvature, so we set $E = 0$ in this scenario. The system (6.13a), (6.13a) is closed by an equation of state $p = w\rho$. To implement a rapid phase transition we will assume that w changes on a time scale much smaller than the background expansion. Thus, the FLRW spacetime is defined by $\{w(t), \rho_0, \Lambda^+\}$.

6.3.3. Specification of the bubble interior

The bubble interior is assumed to be dS in all cases. For simplicity we can pick the flat slicing ($k = 0$) as the evolution in other foliations can be found by the corresponding coordinate transformations (see Appendix C). The evolution of the scale factor is

$$a = \exp\left(H^-(t - t_0)\right), \quad H^- := \sqrt{\Lambda^-/3},$$

so it is completely determined once the vacuum energy density Λ^- inside the bubble is given.

6.3.4. Explicit bubble evolution equations in background coordinates

Once the background solution has been determined we can begin to consider the evolution of the domain wall. Therefore we rewrite $\dot{\bar{R}} = \dot{\bar{t}}\frac{d}{dt}(\bar{a}\bar{r}) = \pm\sqrt{-1 - 2V}$ and solve for $\partial_{\bar{t}}\bar{r}$ to obtain

$$\partial_{\bar{t}}\bar{r} = \frac{-(1 + 2\bar{E})\bar{r}\partial_{\bar{r}}\bar{a} \pm \sqrt{\left(1 + 2\bar{E}\right)(1 + 2V)\left((\bar{r}\partial_{\bar{r}}\bar{a})^2 - 2\bar{E} + 2V\right)}}{(\bar{r}\partial_{\bar{r}}\bar{a} + \bar{a})\left(2\bar{E} - 2V\right)}, \tag{6.14}$$

with V given by equation (6.10). The motion of the domain wall is coupled to the evolution of the surface energy density

$$\partial_{\bar{t}}\sigma = (\bar{\rho} + \bar{p})\frac{(\bar{r}\partial_{\bar{r}}\bar{a} + \bar{a})\partial_{\bar{t}}\bar{r}}{\sqrt{1 + 2\bar{E} - (\bar{r}\partial_{\bar{r}}\bar{a} + \bar{a})^2(\partial_{\bar{t}}\bar{r})^2}}. \tag{6.15}$$

To get a specific solution to these equations we have to provide the areal radius of the bubble $\bar{R}_0$ and its surface energy density σ_0 at some time $\bar{t} = t_0$. This solution can then

6. Spacetime junction

be used to get $\tau(\bar{t})$ and $\bar{t}(\tau)$ upon integration of the matching condition (6.2)

$$d\tau = \sqrt{1 - \frac{(\bar{r}\partial_{\bar{t}}\bar{a} + \bar{a})^2}{1 + 2\bar{E}}(\partial_{\bar{t}}\bar{r})^2}\, d\bar{t}, \tag{6.16}$$

which in turn provides the evolution of the areal radius in terms of the proper time of the domain wall via $\bar{R} = \bar{a}\bar{r}$. This will reveal the effects of inhomogeneity, or a rapid phase transition, in the background on the evolution of $\bar{R}(\tau)$.

At the time $\bar{t} = t_0$, at which we will specify the initial data, the bubble shall be at rest with respect to the exterior background, that is $\partial_{\bar{t}}\bar{r} = 0$ at $\bar{t} = t_0$. This implies

$$\frac{1 + 2\bar{E}}{\bar{a}_0^2 \bar{r}_0^2} = \left(\frac{\varrho}{3\bar{a}_0^3 \sigma_0} + \frac{\Lambda^+ - \Lambda^-}{24\pi\sigma_0} - 2\pi\sigma_0\right)^2. \tag{6.17}$$

Hence, the position and surface energy density of the domain wall at $\bar{t} = t_0$ are not independent. It is also possible to impose $\bar{R} = 0$ at $\bar{t} = t_0$. However, it is more reasonable to require that the bubble is comoving in the parent spacetime at nucleation. This issue does not arise in the pure vacuum case as $\partial_{\bar{t}}\bar{r} = 0$ depends upon slicing and thus $\dot{\bar{R}} = 0$ is the convenient choice.

7. Bubble evolution on dynamical backgrounds

In this chapter we present the analysis of the evolution of a vacuum bubble on backgrounds of different matter content. Therefore we will have to solve the system of equations introduced in the last chapter. We begin with the vacuum case in section 7.1, introduce homogeneously distributed dust in section 7.2 and inhomogeneous dust and curvature in section 7.3. Furthermore we will look at the evolution of the domain wall in a background filled with an homogeneous fluid that undergoes a rapid phase transition in section 7.4.

7.1. Evolution on vacuum background

In a purely vacuum background we have

$$\boxed{p = 0,\ \rho_0 = 0 \Rightarrow \varrho = 0,\ E = 0,\ \Lambda^+ > \Lambda^- > 0.}$$

Of course, $\rho_0 = 0$ means also $\varrho = 0$. The junction spacetimes are both dS but with different vacuum energy densities: $\mathcal{M}^- \equiv \mathrm{dS}_{\Lambda^-}$ and $\mathcal{M}^+ \equiv \mathrm{dS}_{\Lambda^+}$. Thus the two of them can be treated on an equal footing (that is we will omit the indices $\pm$) until we impose the initial data discussed in section 6.3.4. The evolution of the scale factor is

$$a = \exp\left(H\left(t - t_0\right)\right), \quad H := \sqrt{\Lambda/3}.$$

It is normalized such that it equals unity at the time when the initial data of the position and surface energy of the domain wall are specified.

Having established the background dynamics, we can turn to the evolution of the bubble. From equation (6.15) we immediately see that $\sigma = \sigma_0$ is constant, hence the only dynamical variable is the areal radius. The evolution equation for the position of the domain wall (6.14) becomes

$$\partial_t \bar{r} = \frac{-1 \pm u\sqrt{(1 + u^2)\, H^2 \bar{a}^2 \bar{r}^2 - 1}}{(1 + u^2)\, H \bar{a}^2 \bar{r}}, \qquad (7.1)$$

where we have introduced the dimensionless constant

$$u_\pm := \left(\frac{H_+^2 - H_-^2}{8\pi\sigma_0} \mp 2\pi\sigma_0\right) H_\pm^{-1}. \qquad (7.2)$$

7. Evolution on dynamical backgrounds

The geometrical bound (6.8c) implies $u > 0$. Rewriting equation (7.1) as a differential equation for $\bar{a}\bar{r}$ we can solve it by a separation of variables

$$H^2 \bar{r}^2 = \frac{\left(u \pm \sqrt{(1+u^2)H^2 \bar{R}_0^2 - 1}\right)^2}{1+u^2} - \frac{2u\left(u \pm \sqrt{(1+u^2)H^2 \bar{R}_0^2 - 1}\right)}{(1+u^2)\bar{a}} + \frac{1}{\bar{a}^2}. \quad (7.3)$$

Equation (6.16) is difficult to solve already in this case. However, we can make use of the fact that there is no explicit time dependence in the potential

$$2V = -\left(1+u^2\right)H^2 \bar{R}^2, \quad (7.4)$$

to integrate equation (6.7) directly

$$\bar{R} = \frac{\cosh\left(\sqrt{1+u^2}H(\tau - \tau_0)\right)}{\sqrt{1+u^2}H}. \quad (7.5)$$

This is the well known result for the evolution of a spherical domain wall on a vacuum background, it has amply been studied in the literature, see citations in chapter 6. Driven by the pressure difference across the wall the size of the bubble diverges as $\tau \to \pm\infty$.

By equating $\bar{R} = \bar{a}\bar{r}$ this solution immediately yields the map between τ and $\bar{t}$ and thus completes the general solution to the problem.

Under the requirement that the bubble wall is comoving in the flat slicing of dS_{Λ^+} at $\bar{t} = t_0$, that is

$$\boxed{\partial_{\bar{t}} \bar{r} = 0 \big|_{\bar{t}=t_0}},$$

the areal radius becomes

$$\bar{R}_0 = (u_+ H_+)^{-1} = \bar{r}_0 = \left(\frac{H_+^2 - H_-^2}{8\pi\sigma_0} - 2\pi\sigma_0\right)^{-1}, \quad (7.6)$$

in correspondence with equation (6.17). Thus the trajectory of the domain wall in dS_{Λ^+} and dS_{Λ^-} is given by

$$\bar{r}_+ = \sqrt{u_+^{-2} + \left(\bar{a}_+^{-1} - 1\right)^2} H_+^{-1}, \quad (7.7a)$$

$$\bar{r}_- = \sqrt{\frac{1+u^2}{(1+u_+^2)\bar{a}_-^2} - \frac{2u_-(1+u_- u_+)}{u_+(1+u_+^2)\bar{a}_-} + \frac{(1+u_- u_+)^2}{u_+^2(1+u_+^2)}} H_+^{-1}. \quad (7.7b)$$

The trajectory of the domain wall in dS_{Λ^+} is shown in Figure 7.1 for several different surface energy densities. We will not show the trajectories of the domain wall within a causal diagram of dS spacetime, these can be found in Aguirre and Johnson (2005). The reason is that for more general spacetimes, such as LT, no causal diagram can be drawn, cf. Ellis et al. (1985), which prohibits a comparison on this basis.

7.2. Evolution in homogeneous dust

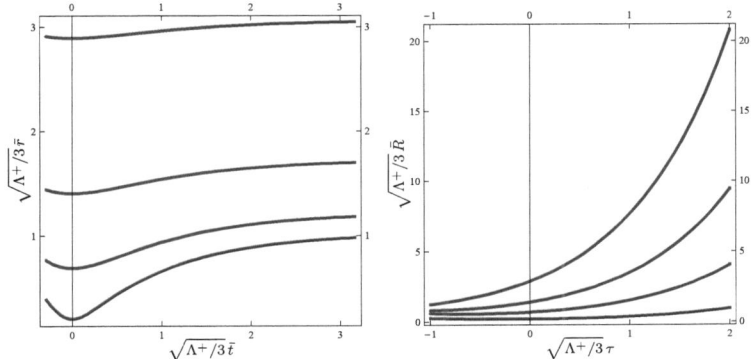

Figure 7.1.: The figure shows the trajectories of $\bar{r}(\bar{t})$ and $\bar{R}(\tau)$ for $\sigma_0 = \{0.1, 0.3, 0.5, 0.7\}\sigma_{max}$. The proper time τ_0 has been chosen such that $\bar{R}(0) = \bar{r}_0$. The domain wall converges to a finite coordinate radius in the dS spacetime, it becomes comoving. Thus, after a while, the physical growth of the bubble is entirely due to the expansion of the background.

7.2. Evolution in homogeneous dust

7.2.1. Evolution in the Einstein Static background

This is the simplest extension to the vacuum case. The Einstein Static (ES) background is a static spacetime. It has constant curvature and dust density and is stabilized by the energy density of the vacuum. For simplicity we will assume that the bubble interior is Minkowski, so we have

$$p = 0, \; \rho_0 = \text{constant}, \; E < 0, \; 4\pi\varrho = \Lambda^+ > 0, \; \Lambda^- = 0.$$

We can set $a = 1$ without loss of generality and thus $\rho_0 = \varrho$ and $\bar{R} = \bar{r}$. The constant positive curvature k determines $E = -k\bar{r}^2$ and also $4\pi\varrho = \Lambda^+ = k$. Since there is no dynamics of the background we can start with the evolution of the bubble right away. The evolution equations of the domain wall are

$$\partial_{\bar{t}}\bar{r} = \pm\sqrt{\frac{(1 - k\bar{r}^2)(-1 - 2V)}{-k\bar{r}^2 - 2V}},$$

where the potential is given by

$$2V = -\left(\frac{\varrho}{3\sigma} + \frac{\Lambda^+}{24\pi\sigma} + 2\pi\sigma\right)^2 \bar{r}^2,$$

7. Evolution on dynamical backgrounds

and the evolution of the surface energy density follows from

$$\partial_{\bar{t}}\sigma = \frac{\varrho \partial_{\bar{t}}\bar{r}}{\sqrt{1 - k\bar{r}^2 - (\partial_{\bar{t}}\bar{r})^2}}.$$

The initial values are parameterized by

$$\sigma_0 = \frac{s_0}{4\pi}\sqrt{\frac{8\pi\varrho}{3} + \frac{\Lambda^+}{3}} = \frac{s_0\sqrt{k}}{4\pi} = s_0 \sigma_{max},$$

with $0 < s_0 < 1$. This translates into an initial radius of

$$k\bar{r}_0^2 = \frac{4s_0^2}{(1+s_0^2)^2}.$$

Though the system looks quite involved, an analytical solution can be obtained by rewriting σ as a function of $\bar{r}$. Therefore we define a function $s(\bar{r})$ through $\sigma = s\sigma_{max}$. This allows to write the evolution equation for σ as

$$2\left(1 - k\bar{r}^2\right)s\partial_{\bar{r}}s + k\bar{r}s^2 - k\bar{r} = 0.$$

This equation is readily solved to give

$$s = \sqrt{1 - (1+s_0^2)\sqrt{1-k\bar{r}^2}}.$$

Hence, we know the surface energy density in terms of $\bar{r}$ and this can be used to obtain the solution for $\bar{r}(\bar{t})$. It reads

$$k\bar{r}^2 = \frac{(1+s_0^2)^2 - \left(2 - (1+s_0^2)\cos\left(\sqrt{k}\bar{t}\right)\right)^2}{(1+s_0^2)^2},$$

and consequently

$$\sigma = \sigma_{max}\sqrt{(1+s_0^2)\cos\left(\sqrt{k}\bar{t}\right) - 1},$$

with $t_0 = 0$. One immediately recognizes that the surface energy density vanishes at

$$t_\pm = \pm\frac{1}{\sqrt{k}}\arccos\frac{1}{1+s_0^2}.$$

However, the size of the domain wall is finite at this time. This suggests to interpret these bubbles as fluctuations in the ES background. Though it is possible to convert the dS solution to a closed coordinate system that has the same spatial geometry as ES, the result is rather involved and does not allow for an easy comparison.

The integral (6.16) can not be solved analytically. Notwithstanding, we note that the integrand is finite within $t_- < \bar{t} < t_+$ and thus the proper time will be finite as well. Therefore the evolution of $\bar{R}(\tau)$ and $\sigma(\tau)$ will resemble that of $\bar{r}(\bar{t})$ and $\sigma(\bar{t})$ respectively, except for the rescaling of the time coordinate.

7.2. Evolution in homogeneous dust

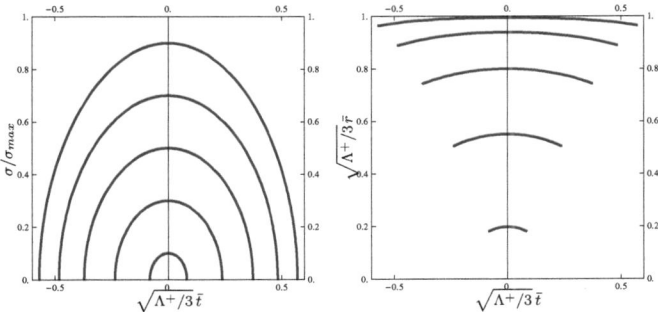

Figure 7.2.: The figure shows the evolution of the surface energy density and the size of the domain wall in the ES spacetime. As this background is static, the time evolution is symmetric. The classical trajectory ends at $\bar{t} = t_\pm$ because the surface energy density vanishes at this time. However, the bubbles have a finite size at this time. This suggests to regard them as fluctuations in the ES background.

7.2.2. Evolution in the Friedmann spacetime

In this section we abandon the static background and proceed with a dynamical dust evolution. In order to obtain an analytical solution of the scale factor, we restrict the consideration to a flat background. That is

$$\boxed{p = 0, \ \rho_0 = \text{constant}, \ E = 0, \ 8\pi\varrho > \Lambda^+ > \Lambda^- > 0.}$$

The equation of motion of the background (6.11a) is readily integrated to give

$$a = a_\star \sinh^{2/3}\left(\frac{\sqrt{3\Lambda^+}}{2}(t - t_B)\right) \quad \text{and} \quad \rho = \frac{\varrho}{a^3},$$

where $a_\star^3 = 8\pi\varrho/\Lambda^+$. The time of the Big Bang t_B follows from the definition of t_0 via $a(t_0, r) = 1$, and is

$$t_B = t_0 - \frac{2}{\sqrt{3\Lambda^+}}\operatorname{arcsinh}\left(a_\star^{-3/2}\right). \tag{7.8}$$

This implies $\rho_0 = \varrho$ and the fact $8\pi\varrho = 10\Lambda^+$ guarantees matter domination at the time when the bubble wall is comoving with the exterior background. We also define the landmark t_1 implicitly by $8\pi\rho(t_1, r) = \Lambda^+$ and analogously t_2 by $8\pi\rho(t_2, r) = 0.1\Lambda^+$. As we study homogeneously distributed dust in this section, these landmarks do not depend on the radial coordinate. This will change in the subsequent sections when radial inhomogeneities are introduced.

7. Evolution on dynamical backgrounds

The background is determined by these equations and we can now have a look at the evolution equation for the position of the domain wall

$$\partial_{\bar{t}}\bar{r} = \frac{-\bar{r}\partial_{\bar{t}}\bar{a} \pm \sqrt{(1+2V)\left((\bar{r}\partial_{\bar{t}}\bar{a})^2 + 2V\right)}}{-2\bar{a}V}, \tag{7.9}$$

where

$$2V = -\left[\frac{\Lambda^-}{3} + \left(\frac{\varrho}{3\bar{a}^3\sigma} + \frac{\Lambda^+ - \Lambda^-}{24\pi\sigma} + 2\pi\sigma\right)^2\right]\bar{a}^2\bar{r}^2. \tag{7.10}$$

and the surface energy density is determined by

$$\partial_{\bar{t}}\sigma = \frac{\varrho\partial_{\bar{t}}\bar{r}}{\bar{a}^2\sqrt{1 - \bar{a}^2\left(\partial_{\bar{t}}\bar{r}\right)^2}}. \tag{7.11}$$

Equation (6.17) provides the position at $\bar{t} = t_0$ by

$$\bar{r}_0 = \left(\frac{\varrho}{3\sigma_0} + \frac{\Lambda^+ - \Lambda^-}{24\pi\sigma_0} - 2\pi\sigma_0\right)^{-1}. \tag{7.12}$$

For a fixed surface energy density the additional dust in the background implies a reduced initial radius of the bubble. This is so, because the energy content of the bubble is completely determined by the energy density of the interior vacuum, the energy density on the surface and its size. This energy has to coincide with the energy that is shed from the exterior background. With the addition of matter to that background the energy budget of the bubble is already reached at a smaller radius. In this way the size of the bubble is determined by the amount of matter density in the background.

The presence of matter has another important consequence as can be seen by writing down the acceleration of the domain wall with respect to the exterior coordinate time. When $\partial_{\bar{t}}\bar{r} = 0$ we have

$$\partial_{\bar{t}}^2\bar{r}\,|_{\partial_{\bar{t}}\bar{r}=0} = \frac{1}{\bar{a}}\left(\frac{\Lambda^+ - \Lambda^-}{24\pi\sigma} - 2\pi\sigma - \frac{2\bar{\rho}}{3\sigma}\right). \tag{7.13}$$

The right hand side is always positive for $\rho = 0$ but it becomes negative when the dust density satisfies

$$\frac{8\pi\bar{\rho}}{\Lambda^+ - \Lambda^-} > \frac{1}{2} - \frac{24\pi^2\sigma^2}{\Lambda^+ - \Lambda^-}.$$

This is illustrated in Figure 7.3 which characterizes the behavior of the domain wall after it was nucleated at rest in the comoving frame of an exterior FLRW spacetime with dust and vacuum energy. With respect to an exterior comoving observer, the bubble shows different behavior in different regions of the σ^2-ρ-plane, which is drawn in units of $\epsilon_{\text{vac}} := (\Lambda^+ - \Lambda^-)/(8\pi)$. The shaded region is forbidden for our choice of junction because σ violates the bound (6.8c). If the energy density of dust ρ is chosen above the dashed red line, the bubble starts to contract. This includes all matter dominated universes $8\pi\rho > \Lambda^+$ since we assume $\Lambda^+ > \Lambda^- > 0$. Below the dashed red line the bubble

7.2. Evolution in homogeneous dust

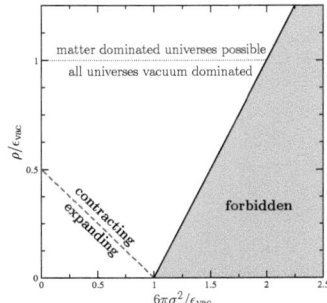

Figure 7.3.: The figure shows the 'acceleration' of the domain wall at times at which it is comoving in the LT spacetime. See text for more exposition.

starts to expand into the ambient spacetime. This includes all de Sitter spacetimes since they are found on the line $\rho = 0$. Below the dotted blue line, all universes are vacuum dominated. As we want to study the evolution of a bubble in a matter dominated background the inequality should always be true at $\bar{t} = t_0$. To be specific, we fix the ratios of the parameters to

$$\boxed{\frac{8\pi\varrho}{\Lambda^+} = 10\,, \quad \frac{\Lambda^-}{\Lambda^+} = 0.1\,,}$$

throughout this work. For completeness we note that we use $\Lambda^+ = 3 \cdot 10^{-5}$ and $t_0 = 0$ in accordance with Fischler et al. (2008), though the actual values are not relevant as the results are normalized by Λ^+. Unfortunately, we can not use any of the simplifications done in the preceeding sections and we must solve the system numerically.

The first new property in the evolution of the domain wall is that the classical trajectory is limited in the past due to the occurrence of the Big Bang. However, the solution breaks down as the surface energy density of the bubble vanishes. Larger bubbles live long enough to experience the rarefaction of the dust and the onset the domination of vacuum energy density. These bubbles then move on a purely vacuum dominated background and thus show the same asymptotic behavior as the solutions in the dS case. The trajectories are shown in Figure 7.4 for several values of the initial surface energy density. For clarity we have included the contour lines of constant dust density.

We proceed with the conversion to a proper time parameterization via equation (6.16) to express the surface energy density and the areal radius of the domain wall in physical time. The result is shown in Figure 7.5. It is seen that the surface energy vanishes at a finite proper time, beyond which the evolution could not be obtained within our approach. However, except for the smallest bubble considered, the surface energy density is well defined in the future of the domain wall's evolution.

7. Evolution on dynamical backgrounds

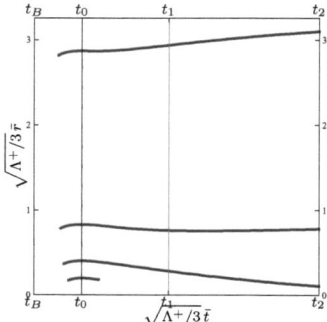

Figure 7.4.: The trajectories of the domain wall in the LT coordinates for several values of the initial surface energy density $\sigma_0 = \{0.3, 0.5, 0.7, 0.9\}\sigma_{max,0}$. The solution breaks down in the past at a time $t_B < \bar{t} < t_0$ but also in the future as the surface energy density of the bubble vanishes.

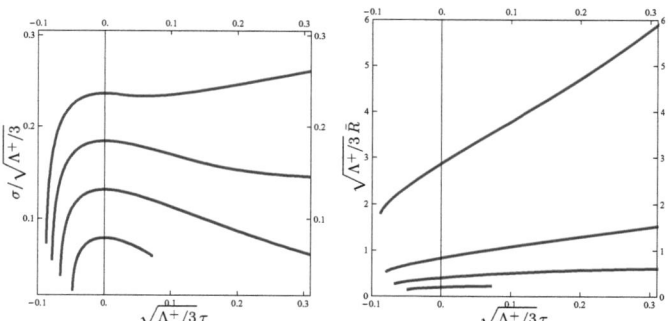

Figure 7.5.: The Figure shows the evolution of the surface energy density and the areal radius in dependence of the proper time of the domain wall for several values of the initial surface energy density $\sigma_0 = \{0.3, 0.5, 0.7, 0.9\}\sigma_{max,0}$. The proper time is chosen such that $\tau = 0$ corresponds to $\bar{t} = t_0$. The solution breaks down when the surface energy density vanishes.

7.3. Evolution through inhomogeneous dust

7.3.1. Evolution without curvature

In this section we will have a look at inhomogeneities in the dust profile in a flat background, such that

$$\boxed{p = 0, \; \partial_r \rho_0 \neq 0, \; E = 0, \; 8\pi\varrho > \Lambda^+ > \Lambda^- > 0.}$$

The background solution can still be found upon analytical integration of equation (6.11a). It resembles the background solution of the homogeneous case

$$a = a_\star \sinh^{2/3}\left(\frac{\sqrt{3\Lambda^+}}{2}(t - t_B)\right) \quad \text{and} \quad \rho = \frac{\varrho}{a^2(r\partial_r a + a)}.$$

The difference is that the inhomogeneous dust profile ρ_0 enforces an inhomogeneous 'scale factor' and an inhomogeneous bang time

$$t_B = t_0 - \frac{2}{\sqrt{3\Lambda^+}}\operatorname{arcsinh}\left(\frac{a_0^{3/2}}{a_\star^{3/2}}\right).$$

We immediately see that $\partial_r \rho_0 > 0 \Leftrightarrow \partial_r t_B > 0$, which reflects the fact that the big bang occurs later where the dust density ρ_0 is larger. The landmarks defined in the last section now also become dependent on the radial coordinate.

As we want to study the evolution of a bubble wall through an ambient matter density we impose the radially increasing dust profile

$$\boxed{\rho_0 = \varrho \exp\left(\sqrt{\Lambda^+/3}\,r - 1\right).} \tag{7.14}$$

The specification of ρ_0 defines a_0 via equation (6.12). This dust profile enforces a shell-crossing singularity at a time $t_B < t < t_0$, as indicated in Figure 7.6. Since our approach is to be understood as toy model of the evolution of a vacuum bubble, beginning with the nucleation in a matter dominated spacetime, we liberally allow the background not to be well defined prior to the nucleation event and cautiously proceed with the analysis.

Having established the background solution we can turn to the evolution of the domain wall. The evolution equation for its position becomes

$$\partial_{\bar{t}} \bar{r} = \frac{-\bar{r}\partial_{\bar{t}}\bar{a} \pm \sqrt{(1 + 2V)\left((\bar{r}\partial_{\bar{t}}\bar{a})^2 + 2V\right)}}{-2(\bar{r}\partial_{\bar{r}}\bar{a} + \bar{a})V},$$

with

$$2V = -\left[\frac{\Lambda^-}{3} + \left(\frac{\varrho}{3\bar{a}^3\sigma} + \frac{\Lambda^+ - \Lambda^-}{24\pi\sigma} + 2\pi\sigma\right)^2\right]\bar{a}^2\bar{r}^2,$$

7. Evolution on dynamical backgrounds

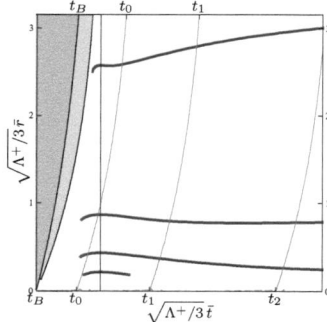

Figure 7.6.: The trajectories of the domain wall in the LT coordinates for several values of the initial surface energy density $\sigma_0 = \{0.3, 0.5, 0.7, 0.9\}\sigma_{max,0}$. In the blue region we have $t < t_B$ and the gray region indicates a negative dust density due to the occurrence of a shell-crossing singularity. The trajectories evolve in the direction of decreasing dust density.

and the evolution of the surface energy density is given by

$$\partial_{\bar{t}}\sigma = \frac{\varrho \partial_{\bar{t}}\bar{r}}{\bar{a}^2\sqrt{1 - (\bar{r}\partial_{\bar{r}}\bar{a} + \bar{a})^2 (\partial_{\bar{t}}\bar{r})^2}}.$$

As before, we assume that the domain wall is comoving with respect to the exterior background at $\bar{t} = t_0$. Hence, the solution to

$$\bar{a}_0 \bar{r}_0 \left(\frac{\varrho}{3\bar{a}_0^3 \sigma_0} + \frac{\Lambda^+ - \Lambda^-}{24\pi\sigma_0} - 2\pi\sigma_0 \right) = 1,$$

determines the radius $\bar{r}_0$. For our choice of the parameters $\varrho, \Lambda^+, \Lambda^-, \sigma_0$ the domain wall is always in a dust dominated environment at $\bar{t} = t_0$, that is $8\pi\rho_0(\bar{r}_0) > \Lambda^+$.

In spite of the radially increasing dust profile, the evolution of the domain wall is very similar to the homogeneous case, as the dust is efficiently diluted by the cosmological expansion. In fact, as can be inferred from Figure 7.6, the domain wall propagates into regions of lower dust density. Moreover, it is seen in Figure 7.7 that the surface energy density sharply decreases as the domain wall comes close to the shell-crossing singularity. However, the evolution of the areal radius is practically not affected by the radially increasing dust density.

7.3. Evolution through inhomogeneous dust

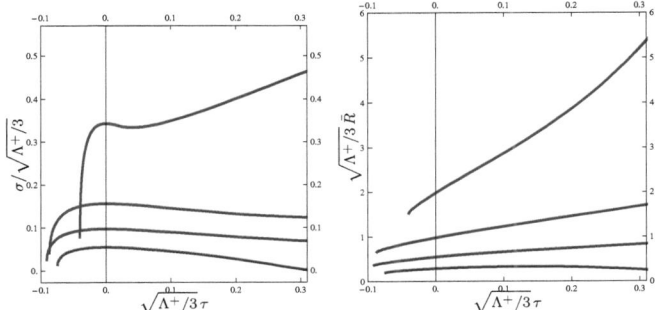

Figure 7.7.: The Figure shows the evolution of the surface energy density and the areal radius in dependence of the proper time of the domain wall for several values of the initial surface energy density $\sigma_0 = \{0.3, 0.5, 0.7, 0.9\}\sigma_{max,0}$. The proper time is chosen such that $\tau = 0$ corresponds to $\bar{t} = t_0$. In contrast to the homogeneous case, the surface energy density sharply decreases when the domain wall is close to the shell-crossing singularity where the dust density becomes infinite.

7.3.2. Evolution with varying curvature

Thus far, we have considered flat backgrounds only. Now we are going to incorporate additional (positive) curvature, such that

$$\boxed{p = 0,\ \partial_r \rho_0 \neq 0,\ \partial_r E \neq 0,\ 8\pi\varrho > \Lambda^+ > \Lambda^- > 0\,.}$$

New properties have to be taken into account when we consider a curved space. Positive curvature means that $E < 0$ and thus possibly leads to a degeneracy of the metric where $1 + 2E = 0$. This will be avoided by a suitable choice of the curvature profile. Furthermore, for a given amount of dust and vacuum energy density there exists a critical amount of curvature that determines whether the space is going to collapse or to expand to infinity. This is familiar from the FLRW spaces, except for the fact that curvature can now depend on the radial coordinate and thus only some parts of the space may collapse while others expand. The scale $\mathcal{R}_{cr}$ that is related to the critical curvature is given by $\mathcal{R}_{cr}^{-3} = 4\pi\varrho\sqrt{\Lambda^+}$. Thus, to avoid a collapse of the background space we have to make sure that $\sqrt{-2E/r^2}\mathcal{R}_{cr} < 1$.

In order to solve the dynamics of the 'scale factor' we take the dust density profile (7.14) from the last section and a curvature profile, for further convenience defined through $k(r) := -2E(r)/r^2$, with

$$\boxed{k = \frac{1}{2\left(\beta\mathcal{R}_{cr}\right)^2}\left(1 + \tanh\left(\frac{r-\delta}{\gamma}\right)\right)\,.}$$

7. Evolution on dynamical backgrounds

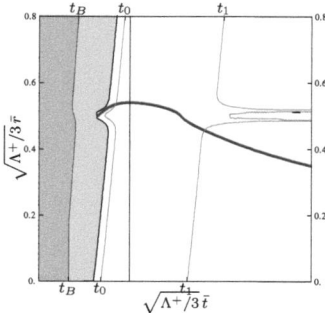

Figure 7.8.: Trajectory of the domain wall in the LT coordinates for $\sigma_0 = 0.7\sigma_{max,0}$. In the blue region we have $t < t_B$ and the gray region indicates a negative dust density due to the occurrence of a shell-crossing singularity. There is a kink in the trajectory as the domain wall runs through the curvature inhomogeneity.

Having determined the background dynamics we will now consider the evolution of the domain wall as given by equations (6.14) and (6.15). The radius at $\bar{t} = t_0$ is to be obtained from the solution to equation (6.17). In the last section we have found that the domain wall covers only a small range in the coordinates of the LT spacetime. Therefore we have to take care when picking the initial surface energy density, because for most choices the bubble will not propagate through the curvature inhomogeneity which is confined to a small region.

In Figure 7.8 we show the trajectory of the domain wall for an initial surface energy density of $\sigma_0 = 0.7\sigma_{max,0}$. It is seen that the curvature inhomogeneity has a sizeable effect on the evolution of the domain wall in the LT coordinates. We will now convert to the proper time parameterization and have a look whether this effect can also be seen in the evolution of the physical radius of the bubble. Therefore we compare the evolution of the domain wall through the LT spacetime with and without curvature inhomogeneities. At nucleation, the physical size of the domain wall shall be the same in both cases, so the initial surface energy densities have to be different. From Figure 7.9 it is seen that the curvature inhomogeneity affects both, the evolution of the surface energy density as well as the areal radius of the domain wall.

7.4. Evolution during a rapid phase transition

In this section the evolution of a vacuum bubble in an homogeneous perfect fluid that undergoes a rapid phase transition will be studied.

$$\boxed{p \neq 0,\ \rho_0 = \text{constant},\ E = 0,\ 8\pi\varrho > \Lambda^+ > \Lambda^- > 0.}$$

7.4. Evolution during a rapid phase transition

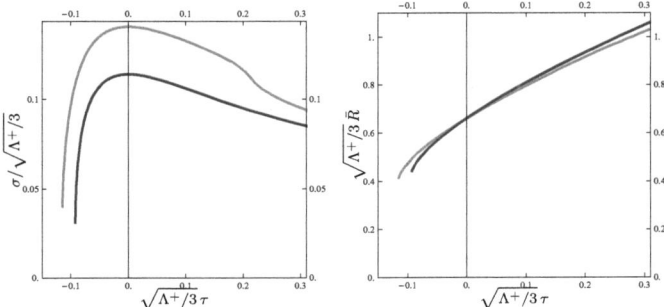

Figure 7.9.: The Figure shows the evolution of the surface energy density and the areal radius in dependence of the proper time of the domain wall for a background with (green) and without curvature inhomogeneity (blue). The proper time is chosen such that $\tau = 0$ corresponds to $\bar{t} = t_0$. The surface densities are chosen such that the domain wall has the same physical radius initially. The curvature inhomogeneity has a sizeable effect on the evolution of the domain wall.

We assume a linear equation of state in the form $p = w\rho$, where we take a time dependent equation of state parameter $w(t)$ to implement the phase transition. In particular, we are interested in the reheating-like transition for which the equation of state parameter changes from $w = -1$ to $w = 1/3$ which motivates to use

$$w = \frac{2}{3} \tanh\left(\frac{t - t_{pt}}{\lambda}\right) - \frac{1}{3}.$$

The timescale λ on which the transition shall occur is taken to be smaller than the timescale $\sqrt{3/\Lambda^+}$ given by the minimal hubble rate. With this profile we numerically solve equations (6.13a) and (6.13b). For convenience we use the same parameters as before and take $\lambda = 0.1\sqrt{3/\Lambda^+}$ and $t_{pt} = 0.3\sqrt{3/\Lambda^+}$. As the equation of state parameter obeys $w < -1/3$ for $t < t_0 = 0$ there is no Big Bang in this model and thus the evolution of the domain wall is studied for $t \geq t_0$ only.

After the background dynamics have been obtained we can go on with the study of the evolution of the domain wall. The equations to be solved numerically are the same as in the homogeneous dust case, i.e. (7.9) with the potential given through (7.10).

In the beginning the background is purely vacuum dominated and the trajectory is therefore very similar to the dS case. As the phase transition occurs the expansion is stopped and the bubble begins to contract in the FLRW coordinates, in turn similar to the homogeneous dust case considered before. With the onset of reheating particles are created in the background the presence of which slows down the physical growth of the

7. Evolution on dynamical backgrounds

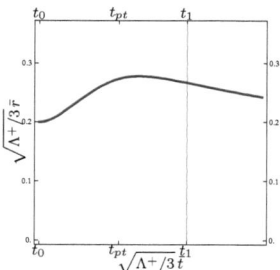

Figure 7.10.: Trajectory of the domain wall in the FLRW coordinates for $\sigma_0 = 0.3\sigma_{max,0}$. The contour line indicates the equality of matter and vacuum energy density.

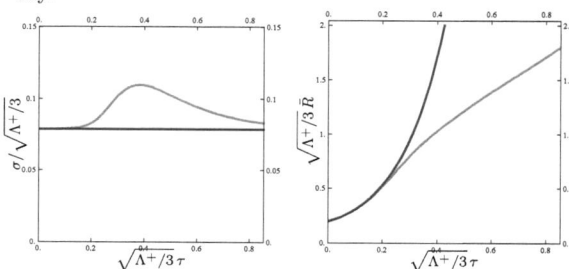

Figure 7.11.: The Figure shows the evolution of the surface energy density and the areal radius in dependence of the proper time of the domain wall for an homogeneous background with (green) and without reheating (blue).

domain wall.

8. Summary and conclusion of Part I

In Part I of this thesis we have studied the evolution of vacuum bubbles on dynamical backgrounds such as the Lemaître-Tolman and Friedmann-Lemaître-Robertson-Walker spacetimes. We have employed a thin shell approach and used the Israel junction method to embed a de Sitter solution into these spacetimes and to track the evolution of the domain wall as determined by the Einstein field equations.

Therefore we started with a review of the relevant properties of the LT spacetime followed by an explicit description of the junction geometry and a derivation of the evolution equations for the background spacetimes and the classical dynamics of the vacuum bubble.

For clarity and to make our work easily accessible and comparable, we have presented the dynamics of vacuum bubbles in a purely vacuum dominated background within our framework. We have extended the scope to consider the evolution of vacuum bubbles in the presence of matter in the surrounding space. We have found that the inclusion of matter allows for several new effects in the evolution of the domain wall.

At first, we have included homogeneously distributed dust in the parent spacetime. The time at which the domain wall is comoving with respect to an observer at rest in that spacetime is referred to as the time of nucleation of the bubble. Though, within this classical approach, the evolution of the domain wall can well be extended to the past of the nucleation event. We have found that the presence of dust affects the size of the domain wall at the time of nucleation. In particular, considering a domain wall of fixed surface energy density, the junction conditions imply that the addition of dust to the background goes along with a decreased areal radius of the bubble at nucleation. This is understood once one recalls that the energy budget of the bubble is required to match the energy that is shed from the background. Given a fixed nucleation radius, the surface energy density of the bubble increases with the amount of dust that is added to the background. Accordingly, with a fixed surface energy density, the size of the bubble becomes the smaller the more dust is present. Our second finding is that the acceleration of the bubble with respect to the LT coordinates becomes negative when sufficient dust is present. For that to happen, the dust must not even be the dominant source of energy in the parent spacetime as illustrated in Figure 7.3. A comoving observer in the LT spacetime, though dominated by vacuum energy density, could see an initial shrinking of the domain wall. This is in contrast to the pure vacuum background where the bubble expands at all times after nucleation in terms of the exterior coordinates. Moreover, we have seen that the junction approach breaks down at a time before to the nucleation event. As the current approach is understood as a toy model for the classical evolution of the bubble *after* it was nucleated in the parent spacetime, this does not mean a problem. However, for some values of the initial surface energy density of the domain wall, the

8. Summary and conclusion of Part I

evolution breakes down at a time to the future of the nucleation event when the shrinking of the domain wall and the expansion of the exterior space causes the surface energy density to vanish. Those bubbles with a larger initial surface energy density continue to expand and as the dust in the background is diluted they resemble the dS solution that was found in the vacuum case.

Furthermore we have studied the evolution of a domain wall in a radially increasing dust profile. Of course, all properties of the homogeneous solution also apply in this case. However, we have shown that even for a dust profile that increases exponentially, the domain wall propagates into regions of lower dust density. This is a vivid illustration of the effectivity of dust rarefaction due to the expansion of the parent space and thus supports the cosmological no hair theorem. This changes when inhomogeneities in the curvature are introduced. The trajectory of the domain wall is significantly affected as it runs through a sharp curvature kink. This is not a mere coordinate effect in the LT spacetime but carries over to a an observer that is comoving with the shell.

We also had a look on the effect of a reheating-like phase transition in the parent spacetime on the evolution of the domain wall. Taking a time dependent equation of state parameter that changes from $w = -1$ to $w = 1/3$ on a timescale much smaller than a hubble time, we have seen that the matter created in the reheating process immediately slows down the expansion of the bubble. The effect is more prominent than the effect due to the varying spatial curvature.

Part II.

Signatures of Bubble Collisions in Minkowski Functionals

9. Introduction to Part II

An inflaton potential with multiple minima allows for the nucleation of bubbles with smaller vacuum energy density. This might occur due to gravitational tunneling via the Coleman-DeLuccia mechanism or via the collision of such bubbles, cf. Easther et al. (2009); Giblin et al. (2010). Pressure difference between the bubble walls force the bubble to expand, quickly approaching the speed of light. In this picture our observable universe can be thought of as a bubble residing among a multitude of bubbles in a 'multiverse'.

However, bubbles inevitably collide and recent work shows that such collisions may leave observable imprints in the CMB. A generic prediction is that a past collision on our bubble universe by another bubble will leave a cold or hot disk – regions where the mean temperature is statistically different - on the CMB sky Aguirre et al. (2007); Aguirre and Johnson (2008); Aguirre et al. (2009); Chang et al. (2008); Chang et al. (2009). In addition to such shift in mean temperatures, the CMB may exhibit additional polarization modes in such regions, cf. Czech et al. (2010), and perhaps lead to anisotropic large scale galaxy flows, see Larjo and Levi (2010). There have been some claims in the literature, by Cruz et al. (2005, 2006, 2007), regarding the existence of such a spot, the so-called "cold spot" in the CMB, using wavelet analysis, although such claims have been challenged by Zhang and Huterer (2010); Bennett et al. (2010). More recently, a model independent pipeline was constructed by Feeney et al. (2010a,b) to search for such signals using causal boundaries and found several possible hints of such features. In this work we will attempt to search for such a signal using a different statistic – Minkowski Functionals (MFs).

Since spots in an otherwise gaussian sky are topological in nature, this suggests the use of statistical descriptors which are well suited to quantify morphological properties of the temperature fluctuations. MFs are exactly such tools – they are morphological statistics on smooth maps. While they are widely used in image processing (e.g. Mantz et al. (2008)) as such, they were first used by cosmologists to look from deviations from Gaussianitiy of the perturbations in large scale structure Schmalzing et al. (1996a,b); Hikage et al. (2003) and the CMB Winitzki and Kosowsky (1998); Schmalzing and Górski (1998); Novikov et al. (1999); Eriksen et al. (2004); Hikage et al. (2006, 2008, 2009); Komatsu et al. (2009b); Matsubara (2010).

In this work we apply MFs to the search for disk-like structures in the CMB – structures expected if our present "bubble universe" has had the (mis)-fortune of colliding with another bubble in the distant past. In chapter 10 we introduce MFs for scalar fields on the sphere. This is followed by an introduction to the HEALPix software package and our implementation of the MF algorithm in chapter 11. In particular we derive analytical formulas to remove numerical "residuals" introduced by binning and derive a general map-independent residual-free estimator. In chapter 12 we present the application of our

9. Introduction to Part II

Figure 9.1.: Simulated CMB maps containing a disk with angular diameter $40°$ and temperature difference $\delta T = 2\sqrt{\sigma_G}$ (upper panel) and $\delta T = \sqrt{\sigma_G}$ (lower panel).

algorithm to Gaussian and collision maps. Therefore we review the standard results for Gaussian random fluctuations, derive the respective residuals and quantify the remaining error in section 12.1. In Section 12.2 we argue what MFs are to be expected in the presence of a bubble collision and analyze their sensitivity. Moreover, we propose our analytical model for disks in the CMB and apply our estimator to the constraint of both simulated collisions maps and the WMAP7 data. In chapter 13 we summarize and conclude.

The content partly follows Lim and Simon (2011) in which the work presented here has been published.

10. Minkowski Functionals on the two-sphere

10.1. Definition

MFs are a tool to characterize the morphological properties of convex, compact sets in an n dimensional space. A property is considered to be morphological when it is invariant under rigid motions, i.e. translations and rotations. Hadwinger's Theorem in integral geometry states that under simple requirements any morphological property can be expanded as a linear combination of $n+1$ functionals, the so-called Minkowski Functionals. We will give a definition of the MFs and briefly introduce their use for characterizing smooth scalar fields on $\mathbb{S}^2$, e.g. temperature anisotropies.

On $\mathbb{S}^2$ there are three MFs which, up to normalization, represent the volume, circumference and integrated geodesic curvature of a given excursion set. The excursion set and the boundary of the excursion set of a smooth scalar field u at threshold ν are defined by

$$Q_\nu := \left\{ x \in \mathbb{S}^2 \mid u(x) > \nu \right\},$$
$$\partial Q_\nu := \left\{ x \in \mathbb{S}^2 \mid u(x) = \nu \right\}.$$

The first Minkowski functional $v_0(\nu)$ is the area fraction of Q_ν. It is given by

$$v_0(\nu) := \frac{1}{4\pi} \int_{\mathbb{S}^2} d\Omega \, \Theta(u - \nu), \qquad (10.2)$$

where Θ is the Heaviside step function. Strictly speaking we use the surface densities of the MFs which are the honest MFs divided by 4π. We consistently use the term MFs for these densities. The second Minkowski functional is proportional to the total boundary length of Q_ν

$$v_1(\nu) := \frac{1}{16\pi} \int_{\partial Q_\nu} dl = \frac{1}{16\pi} \int_{\mathbb{S}^2} d\Omega \, \delta(u - \nu) \, |\nabla u|. \qquad (10.3)$$

Here δ is the Delta distribution and $|\nabla u|$ is the norm of the gradient of u. Finally, the third Minkowski functional is the integral of the geodesic curvature κ along the boundary

$$v_2(\nu) := \frac{1}{8\pi^2} \int_{\partial Q_\nu} dl \, \kappa = \frac{1}{8\pi^2} \int_{\mathbb{S}^2} d\Omega \, \delta(u - \nu) \, |\nabla u| \, \kappa. \qquad (10.4)$$

Geodesic curvature is the quantitative measure of how much the boundary curve γ deviates from being geodetic. For a unit speed curve, i.e. $|\dot{\gamma}| = 1$, it is defined through

$$\kappa := |\nabla_{\dot{\gamma}} \dot{\gamma}|, \qquad (10.5)$$

10. Minkowski Functionals on the two-sphere

where $\nabla_{\dot\gamma}$ represents the covariant derivative along the tangent vector $\dot\gamma$ of the curve. Thus κ vanishes if and only if γ is geodetic as follows from the geodesic equation $\nabla_{\dot\gamma}\dot\gamma = 0$.

For the actual calculation of v_2 we have to express κ in terms of u. To do so, we use the fact that u does not change along γ and thus $du(\dot\gamma) = 0$ which implies that $\dot\gamma^\mu = \epsilon^{\mu\nu}\nabla_\nu u$. We can now choose to either normalize $\dot\gamma$ and use that in equation (10.5), or keep $\dot\gamma$ and rewrite equation (10.5) for a non-unit speed curve. We take the latter approach and note that for a non-unit speed curve equation (10.5) becomes

$$\kappa = |\dot\gamma|^{-3}\sqrt{|\nabla_{\dot\gamma}\dot\gamma|^2 |\dot\gamma|^2 - (\dot\gamma\cdot\nabla_{\dot\gamma}\dot\gamma)^2}\,.$$

The individual terms are

$$|\dot\gamma|^2 = \sin^2\vartheta\left(u_{;\vartheta}^2 + u_{;\varphi}^2\right),$$

$$\dot\gamma\cdot\nabla_{\dot\gamma}\dot\gamma = \sin^3\vartheta\left[u_{;\vartheta}u_{;\varphi}\left(u_{;\vartheta\vartheta} - u_{;\varphi\varphi}\right) - \left(u_{;\vartheta}^2 - u_{;\varphi}^2\right)u_{;\vartheta\varphi}\right],$$

$$|\nabla_{\dot\gamma}\dot\gamma|^2 = \sin^4\vartheta\left[(u_{;\vartheta}u_{;\varphi\varphi} - u_{;\varphi}u_{;\vartheta\varphi})^2 + (u_{;\varphi}u_{;\vartheta\vartheta} - u_{;\vartheta}u_{;\vartheta\varphi})^2\right].$$

The shorthand semicolon notation denotes derivatives of u defined by

$$u_{;\vartheta} := \partial_\vartheta u,\quad u_{;\varphi} := \frac{1}{\sin\vartheta}\partial_\varphi u,\quad u_{;\varphi\varphi} := \frac{1}{\sin^2\vartheta}\partial_\varphi^2 u + \frac{\cos\vartheta}{\sin\vartheta}\partial_\vartheta u,$$
$$u_{;\vartheta\vartheta} := \partial_\vartheta^2 u,\quad u_{;\vartheta\varphi} := \frac{1}{\sin\vartheta}\partial_\vartheta\partial_\varphi u - \frac{\cos\vartheta}{\sin^2\vartheta}\partial_\varphi u = u_{;\varphi\vartheta}\,. \qquad(10.7)$$

Consequently, the geodesic curvature reduces to

$$\kappa = \frac{2u_{;\vartheta}u_{;\varphi}u_{;\vartheta\varphi} - u_{;\vartheta}^2 u_{;\varphi\varphi} - u_{;\varphi}^2 u_{;\vartheta\vartheta}}{\left(u_{;\vartheta}^2 + u_{;\varphi}^2\right)^{3/2}}\,. \qquad(10.8)$$

This completes the necessary formulae to compute the MFs for a given field u.

11. Application of the HEALPix software package

11.1. Basic introduction to HEALPix

HEALPix is the acronym for Hierarchical Equal Area isoLatitude Pixelization that was introduced by Górski et al. (2005). The name refers to a tessellation of the two-dimensional spherical surface with pixels that have identical surface areas, the centers of which are, by definition, located on rings of constant latitude. At base resolution HEALPix consists of 12 pixels that are aligned along three circles at the poles and the equator. The pixels are curvilinear quadrilaterals that can be mapped to the square $[0,1] \times [0,1]$ which allows to obtain a straightforward subdivision of each into $N_{\text{side}} \times N_{\text{side}}$ smaller elements. Accordingly, N_{side} specifies the resolution by the total number of pixels $N_{\text{pix}} = 12 N_{\text{side}}^2$ meaning that each pixel has a surface area of $\Omega_{\text{pix}} = \pi / (3 N_{\text{side}}^2)$. The pixel centers are defined on $4 N_{\text{side}} - 1$ isolatitude rings, equidistant in azimuth on each individual ring. In the equatorial belt (i.e. $|\cos \vartheta| < 2/3$) there are $4 N_{\text{side}}$ pixel centers distributed on each ring, while on the polar cap regions ($|\cos \vartheta| \geq 2/3$) the number of pixels varies, decreasing from ring to ring as distance to the pole decreases. The HEALPix tessellation and resolution are illustrated in Figure 11.1.

The HEALPix software package contains a large set of programs. We shall briefly introduce those that were employed in the course of this work. The command `input_map` assigns field values to each pixel as provided by an input FITS (Flexible Image Transport System) file. The FITS data format has become standard for the storage of astronomical data, in particular the LAMBDA (Legacy Archive Microwave Background Data Analysis) temperature and polarization maps of the CMB come in this data format. The further analysis of the map relies on its spectrum in a spherical harmonics expansion. The corresponding a_{lm}'s are computed via `map2alm` by FFT (Fast Fourier Transformation) up to a cutoff l_{max} that is usually related to the resolution by $l_{\text{max}} = 3 N_{\text{side}}$, to ensure $\sqrt{\Omega_{\text{pix}}} \simeq \pi / l_{\text{max}}$. Moreover, HEALPix provides the subroutine `create_alm` that allows to generate a set of Gaussian a_{lm}'s in accordance with a given power spectrum. In addition, a Gaussian smoothing kernel may be applied to a set of a_{lm}'s through `alter_alm`. A given set of a_{lm}'s allows for an assignment of a field value to each pixel.

The HEALPix software as well as a complete documentation can be found at the HEALPix webpage http://healpix.jpl.nasa.gov/.

11. Application of the HEALPix software

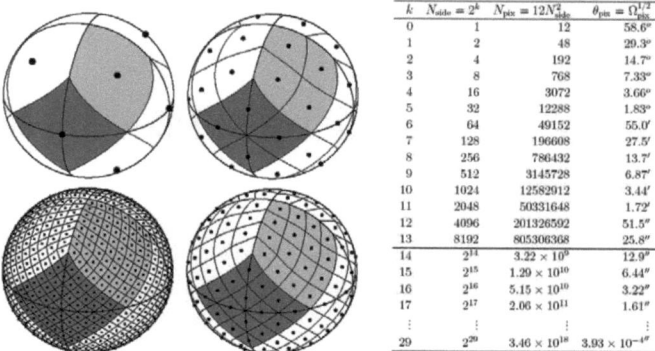

Figure 11.1.: The left panel shows the HEALPix tessellation of the sphere for $N_{\text{side}} = 1, 2, 4, 8$. The table on the right hand side summarizes the relationship between N_{side} and the corresponding total number of pixels and angular resolution. Taken from Górski et al. (2005).

11.2. Extraction of Minkowski Functionals

Though the HEALPix software provides a large number of programs, it does not contain a procedure to compute the MFs of a given map and therefore such a procedure had to implemented. Our approach for the extraction of MFs is a straightforward numerical calculation of the integrals in equations (10.2)-(10.4) as prescribed in Schmalzing and Górski (1998). Given a map, the numerical Minkowski Functional V_i is obtained via a sum of the respective integrand

$$\mathcal{I}_0(\nu, x_k) := \Theta(u - \nu),$$
$$\mathcal{I}_1(\nu, x_k) := \frac{1}{4}\delta(u - \nu)\sqrt{u_{;\vartheta}^2 + u_{;\varphi}^2}, \tag{11.1a}$$
$$\mathcal{I}_2(\nu, x_k) := \frac{1}{2\pi}\delta(u - \nu)\frac{2u_{;\vartheta}u_{;\varphi}u_{;\vartheta\varphi} - u_{;\vartheta}^2 u_{;\varphi\varphi} - u_{;\varphi}^2 u_{;\vartheta\vartheta}}{u_{;\vartheta}^2 + u_{;\varphi}^2}, \tag{11.1b}$$

over all pixels

$$V_i(\nu) := \frac{1}{N_{\text{pix}}} \sum_{k=1}^{N_{\text{pix}}} \mathcal{I}_i(\nu, x_k).$$

The integrands $\mathcal{I}_1$ and $\mathcal{I}_2$ involve the functions $u_{;\mu\nu}$ which represent the shorthand notation for a combination of partial derivatives of u as defined in equations (10.7). The partial derivatives of the temperature field are calculated in Fourier space so that

$\partial_\varphi Y_{lm} = im Y_{lm}$ and

$$\partial_\vartheta Y_{lm} = \begin{cases} \frac{l}{\tan\vartheta} Y_{lm} - \sqrt{\frac{2l+1}{2l-1}} \frac{\sqrt{l^2-m^2}}{\sin\vartheta} Y_{l-1m}, & |m| < l, \\ \frac{l}{\tan\vartheta} Y_{lm}, & |m| = l. \end{cases}$$

The second derivatives can be obtained from these equations in a straightforward way.

Moreover, the delta function that appears in equations (11.1a) and (11.1b) is numerically approximated through a discretization of threshold space in bins of width $\Delta\nu$ by

$$\delta_N(x) := (\Delta\nu)^{-1} \left[\Theta\left(x + \Delta\nu/2\right) - \Theta\left(x - \Delta\nu/2\right)\right].$$

To quantify the numerical error of this approximation notice that if we replace the delta function $\delta(u-\nu)$ in the integral

$$v_i(\nu) = \int_{-\infty}^{\infty} du\, \delta(u-\nu) v_i(u),$$

with the numerical delta function $\delta_N(u-\nu)$ we get

$$V_i(\nu) = v_i(\nu) + R_i^{\Delta\nu}(\nu),$$

with residuals defined through

$$R_i^{\Delta\nu}(\nu) := \left[\frac{1}{\Delta\nu} \int_{\nu-\Delta\nu/2}^{\nu+\Delta\nu/2} du\, v_i(u)\right] - v_i(\nu),$$

As it does not depend on the actual functional form of v_i, this is a general result and represents a generic map-independent (but binsize dependent) MF estimator.

The following example is intended to illustrate our implementation by applying it to a presumably familiar map: Earth's topography.

Example: MFs of earth's topography

The National Geophysical Data Center provides elevation data of the earth's surface in the ETOPO5 data set (NOAA (1988)). We have binned the data set into a HEALPix grid of resolution $N_{\text{side}} = 512$, smoothed it with a Gaussian filter of $1°$ and computed the MFs as prescribed above. The results are shown in Figure 11.3. The functionals reveal a sharp change between 6000m and 5000m below sea-level which means that a large fraction of the sea-floor is to be found at these depths, cf. the average depth of the sea-floor $\simeq 3790$m. The following broad peak in V_1 at $\simeq -4000$m can be associated with the rise of the oceanic ridges. For larger thresholds, the roughly constant boundary length between -3000m and slightly positive elevations indicates the appearance of the continental landmasses. Elevations larger than 1000m make up only a small fraction in earth's topography so that the MFs begin to fade for larger thresholds.

11. Application of the HEALPix software

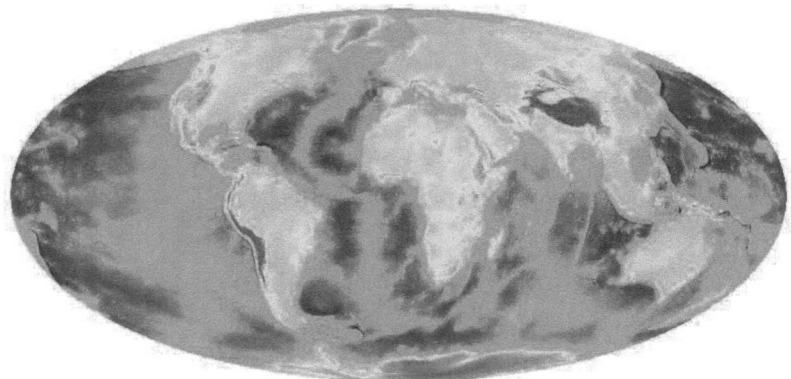

Figure 11.2.: Mollweide projection of Earth's topography. This map was constructed by binning the ETOPO5 data (NOAA (1988)) into a HEALPix grid with resolution $N_{\text{side}} = 512$.

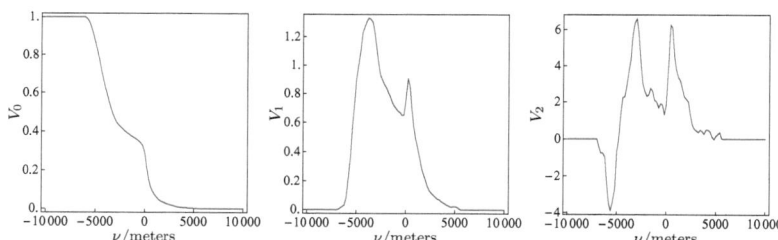

Figure 11.3.: The figure shows the MFs of the ETOPO5 data set of earth's topography with 1° Gaussian smoothing. Prominent features in the functionals can be associated with characteristic topographical properties: oceanic ridges, continents, mountain ranges.

12. Minkowski Functional statistics of a collision signal

12.1. Gaussian random field

For a Gaussian random field u_G one can compute the expectation values of the integrals in equations (10.2) – (10.4). The derivation is relegated to appendix E. The results are

$$\bar{v}_0^G(\nu) := \left\langle v_0^G(\nu) \right\rangle = \frac{1}{2}\left(1 - \mathrm{erf}\left(\frac{\nu - \mu}{\sqrt{2}\sigma}\right)\right), \tag{12.1a}$$

$$\bar{v}_1^G(\nu) := \left\langle v_1^G(\nu) \right\rangle = \frac{1}{8}\sqrt{\frac{\tau}{\sigma}}\exp\left(-\frac{(\nu - \mu)^2}{2\sigma}\right), \tag{12.1b}$$

$$\bar{v}_2^G(\nu) := \left\langle v_2^G(\nu) \right\rangle = \frac{1}{(2\pi)^{3/2}}\frac{\tau}{\sigma}\frac{\nu - \mu}{\sqrt{\sigma}}\exp\left(-\frac{(\nu - \mu)^2}{2\sigma}\right), \tag{12.1c}$$

where erf is the Gaussian error function $\mathrm{erf}(x) := \frac{2}{\sqrt{\pi}}\int_0^x dt\, \exp\left(-t^2\right)$ and

$$\mu := \langle u_G \rangle, \quad \sigma := \langle u_G^2 \rangle - \mu^2, \quad \tau := \frac{1}{2}\left\langle |\nabla u_G|^2 \right\rangle. \tag{12.2}$$

The variance σ and the mean amplitude of the gradient are directly related to the Gaussian angular power spectrum C_l by

$$\sigma = \frac{1}{4\pi}\sum_{l=1}^{\infty}(2l+1)C_l, \quad \tau = \frac{1}{8\pi}\sum_{l=1}^{\infty}(2l+1)(l+1)lC_l.$$

However, note CMB power spectra are often truncated at large l – hence it is preferable to compute these quantities directly from the maps according to equation (12.2).

Assuming that the primordial spectrum is completely Gaussian and the power spectrum is isotropic, we pose the question: *If a "cold" or "hot" spot exists in the CMB (whatever the origin) – how well can we distinguish such a spot from the complete gaussian sky with MF?* By "hot" or "cold" spot, we mean a circular region of size A in the CMB with a uniform temperature shift δT over the *actual* mean temperature $\mu_G = \mu - A/(4\pi)\delta T$ and the usual power spectrum of anisotropies, where μ is the average temperature of the CMB, and we have assumed a sharp cut-off at the boundary.

Furthermore, the *actual* variance of the unaffected region of the sky σ_G is related to the disk properties by $\sigma_G = \sigma - A/(4\pi)\left(1 - A/(4\pi)\right)\delta T^2$, where σ is the variance of the whole sky calculated assuming that no such disk exist. In general, the "hot" or

12. Minkowski Functional statistics of a collision signal

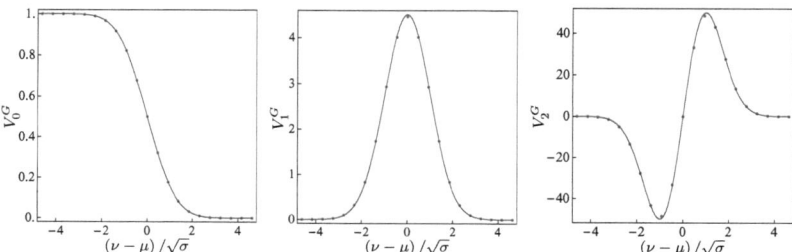

Figure 12.1.: Average numerical MFs V_i^G for a Gaussian random field generated from the power spectrum derived from the five year WMAP data at $N_{\text{side}} = 512$ with $\vartheta_s = 1°$ smoothing (red dots), compared to the expectation value $\bar{v}_i^G$ (blue line) as given in eqs. (12.1a)-(12.1c).

"cold" spot can be fairly complicated in structure, with gentler boundaries and non-uniform temperature shift. In this work, we will ignore such complications and consider a single sharp boundary region with uniform temperature shift. Chang et al. (2009) have predicted such regions by studies of cosmological bubble collisions.

In addition, we would like to point out some subtleties when using MF to test for the presence of disk-like structures. First, MF "know nothing" about the structure of the two-point correlation function, so one in principle can have a completely gaussian map, yet is somehow correlated pixel by pixel – imagine for example, an omnipotent hand rearranging all the cold pixels in a gaussian map into a disk, then we will not pick up such a feature. Since we are assuming that the underlying power spectrum is isotropic, such magical rearranging does not occur. This illustrates one of the advantage of using MF over "local" statistics – we will not mistake fortuitous (but random) correlations as a true feature.

Second, imposing a gaussian disk with different mean on the sky, effectively renders our map to become bi-distributional, i.e. a histogram of the pixels will reveal two different gaussian distributions with different means but identical variances, zero skewness and kurtosis. Hence the sky becomes non-Gaussian. Nevertheless, this is a very specific form of *anisotropic* non-Gaussianity, which cannot be described by a regular higher-point correlation function such as the bispectrum or trispectrum, see Matsubara (2010), which is *isotropic* by construction. Thus, in principle we can hope to distinguish such anisotropic non-Gaussianities from those generated during primordial inflation, cf. Linde and Mukhanov (1997); Bernardeau and Uzan (2003); Maldacena (2003); Chen (2010). We will show later that they possess a very distinctive signature.

Figure 12.1 shows a comparison of the numerical MFs with the expectation values (12.1a-12.1c).

12.1. Gaussian random field

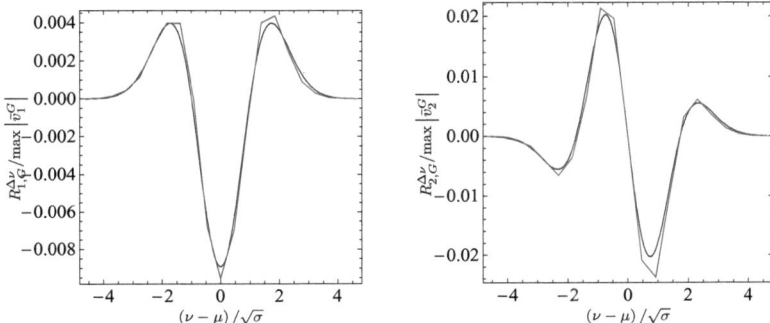

Figure 12.2.: The figure shows the average difference $\left\langle \Delta_i^G \right\rangle$ (red) of the numerical and the analytical MFs and the leading order residual $R_{i,G}^{\Delta\nu}$ with $\Delta\nu \simeq 0.46\sqrt{\sigma}$ (blue) for an average over 256 realizations at $N_{\text{side}} = 512$ and $\vartheta_s = 1°$ normalized to $\max\left|\bar{v}_i^G\right|$. The standard deviation of these maps is $\sqrt{\sigma} \simeq 0.065\text{mK}$.

12.1.1. Removal of residuals

Regarding the residuals we expect that the numerical computation of the MFs to leading order around $\Delta\nu = 0$ yields,

$$V_i^G(\nu) \simeq \bar{v}_i^G(\nu) + \frac{(\Delta\nu)^2}{24}\partial_\nu^2 \bar{v}_i^G(\nu), \quad i \in \{1,2\}.$$

For convenience we define

$$R_{i,G}^{\Delta\nu}(\nu) := \frac{(\Delta\nu)^2}{24}\partial_\nu^2 \bar{v}_i^G(\nu), \quad R_{0,G}^{\Delta\nu} := 0,$$

which means that

$$R_{1,G}^{\Delta\nu}(\nu) = \frac{(\Delta\nu)^2}{24\sigma}\left(\frac{(\nu-\mu)^2}{\sigma} - 1\right)\bar{v}_1^G(\nu), \tag{12.3a}$$

$$R_{2,G}^{\Delta\nu}(\nu) = \frac{(\Delta\nu)^2}{24\sigma}\left(\frac{(\nu-\mu)^2}{\sigma} - 3\right)\bar{v}_2^G(\nu). \tag{12.3b}$$

The average of the difference

$$\Delta_i^G(\nu) := V_i^G(\nu,\mu,\sigma,\tau) - \bar{v}_i^G(\nu,\mu,\sigma,\tau), \quad i \in \{1,2\}, \tag{12.4}$$

between the numerically extracted MFs V_i^G and the respective expectation value $\bar{v}_i^G$, as shown in Figure 12.2, is in very good agreement with $R_{i,G}^{\Delta\nu}$ when a sufficiently large

12. Minkowski Functional statistics of a collision signal

number of realizations is considered. All maps are corrected for the numerical fluctuation in the mean such that $\mu = \mathcal{O}\left(10^{-18}\right)$. The difference Δ_i^G is normalized to the maximal amplitude in the MFs

$$\max\left|\bar{v}_0^G\right| = 1, \quad \max\left|\bar{v}_1^G\right| = \frac{1}{8}\sqrt{\frac{\tau}{\sigma}}, \quad \max\left|\bar{v}_2^G\right| = \frac{1}{(2\pi)^{3/2}}\frac{1}{\sqrt{e}}\frac{\tau}{\sigma}.$$

As we emphasized previously, this calculation can be done for any underlying smooth map.

An expansion of eqns. (12.3a) and (12.3b) around $\Delta\nu = 0$ shows that the leading order terms are proportional to $(\Delta\nu)^2/\sigma$. This fact may suggest that a smaller binsize is always better. However, smaller binsize means that each bin contains fewer pixels and hence an increase in the inherent noise per bin. In Figure 12.3 we show that, for a single realization, a binsize of $\Delta\nu = 0.9\sqrt{\sigma}$ is a good compromise. Another way of beating down the noise is to increase the number of pixels, either by increasing the resolution of the map, or average over a large sample of maps. Figure 12.4 shows a comparison of the residuals $R_{i,G}^{\Delta\nu}$ with the difference Δ_i^G at the prospective Planck resolution $N_{\text{side}} = 2048$ for a single sample at different binsizes. With the number of pixels per bin increased a smaller binsize can be used without adding noise.

However, as we would want to apply our prescription to actual data, we ultimately want to be able to extract accurate MF from a single map. As we shall see, the noise from a single map will turn out to be a difficult stumbling block in our attempt to constrain disk-like structures in the sky.

12.1.2. Remaining difference in MFs of Gaussian maps

The residual effects that originate in the numerical implementation of the delta function have been analyzed in detail in the last subsection. Henceforth, we will use a suitable binwidth in the calculation of MFS for any given map to obtain well converged residuals. Their subtraction then allows for an efficient removal of the effects of the discrete delta function. Consequently we are interested in further effects that may cause a difference between the numerical MFs and their analytical expectation. Therefore we investigate the difference

$$\Delta_i^G(\nu) := V_i(\nu) - \left[\bar{v}_i^G(\nu) + R_{i,G}^{\Delta\nu}(\nu)\right], \tag{12.5}$$

that remains after the residuals have been removed. Examples of this remaining difference in the MFs of one realization of the Gaussian field are shown in Figure 12.5, while the average of the differences are shown in Figure 12.6. The difference in the MFs of a particular realization differs from sample to sample and therefore has a random character.

Moreover, for most samples the shape of the difference Δ_i^G appears to be dominated by $\pm\left(\sqrt{\sigma}\partial_\nu\right)^3 \bar{v}_i^G$ times some random prefactor of the order $(\mathcal{O}(0.1))^3$. The averages of the difference, $\left\langle\Delta_i^G\right\rangle$, converge to a curve that is approximately given by $(\mathcal{O}(0.1)\sqrt{\sigma}\partial_\nu)^4 \bar{v}_i^G$ and are thus much smaller than the random fluctuation in the MFs of a single sample.

12.1. Gaussian random field

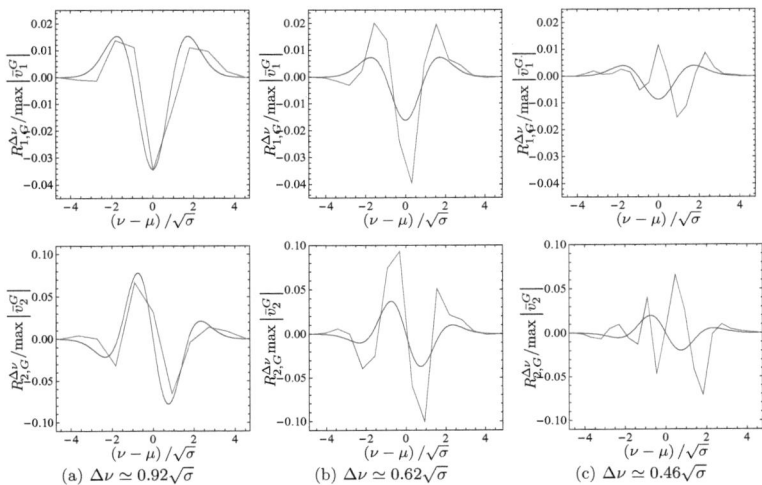

Figure 12.3.: The figure shows the difference Δ_i^G (red) of the numerical and the analytical MFs and the leading order residual $R_{i,G}^{\Delta\nu}$ (blue) for a single realization taken at $N_{\text{side}} = 512$ smoothed to $\vartheta_s = 1°$ at the binwidths $\Delta\nu \simeq (0.92\sqrt{\sigma}, 0.62\sqrt{\sigma}, 0.46\sqrt{\sigma})$ normalized to $\max \left|\bar{v}_i^G\right|$. The upper (lower) panel is Δ_1^G (Δ_2^G).

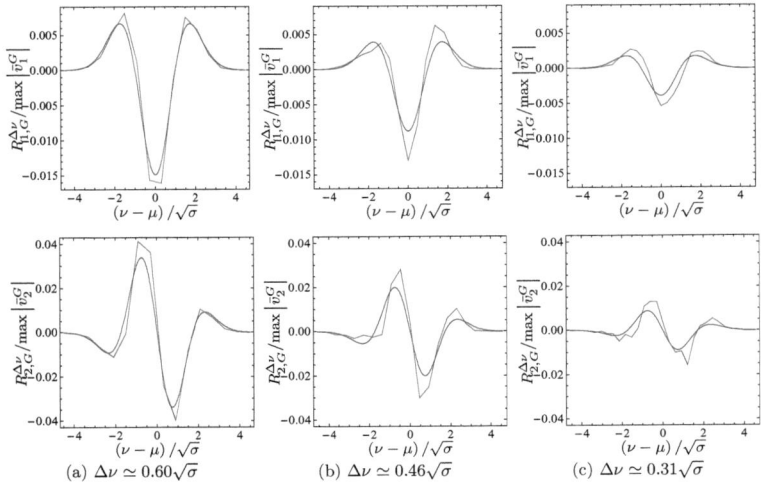

Figure 12.4.: Same as above with $N_{\text{side}} = 2048$ and without smoothing. With increased number of pixels that comes with higher resolution, we can use smaller binsizes while keeping the pixel noise tolerances manageable.

63

12. Minkowski Functional statistics of a collision signal

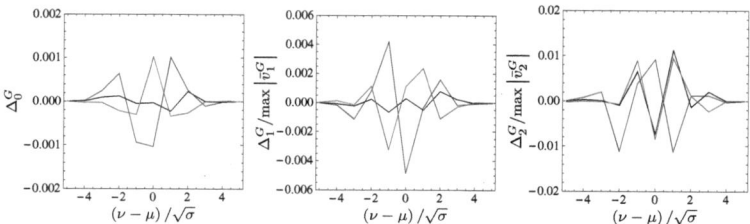

Figure 12.5.: The figure shows the difference Δ_i^G, eq. (12.5), of the numerical and the analytical MFs with residuals subtracted for three different realizations taken at $N_{\text{side}} = 512$ without smoothing at the binwidth $\Delta\nu/\sqrt{\sigma} = 1$ and normalized to $\max\left|\bar{v}_i^G\right|$. Though the differences Δ_i^G are different in each sample they appear to be dominated by $(\mathcal{O}(0.1)\sqrt{\sigma}\partial_\nu)^3 \bar{v}_i^G$.

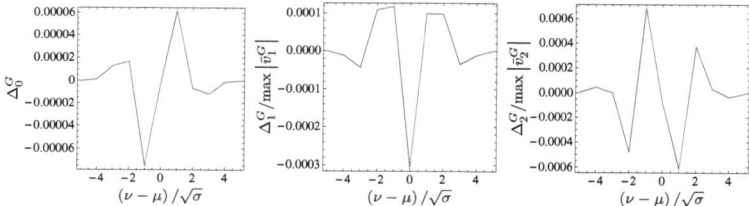

Figure 12.6.: The figure shows the difference Δ_i^G, eq. (12.5), of the numerical and the analytical MFs with residuals subtracted, averaged over 1000 realizations taken at $N_{\text{side}} = 512$ without smoothing at the binwidth $\Delta\nu/\sqrt{\sigma} = 1$ and normalized to $\max\left|\bar{v}_i^G\right|$. The averages $\left\langle\Delta_i^G\right\rangle$ appear to converge to a curve that is approximately given by $(\mathcal{O}(0.1)\sqrt{\sigma}\partial_\nu)^4 \bar{v}_i^G$. However, when compared with the upper figure, it turns out that the fluctuations for a single realization are about an order of magnitude larger than the average.

We point out that this difference in a single sample depends on the resolution (though only weakly $\propto N_{\text{side}}^{-1/3}$) and does not depend on the binwidth $\Delta\nu$.

12.2. Analysis of collision maps

12.2.1. Expected disk signal in MFs

In this section, we describe our method of constructing a sample gaussian map with a superimposed disk. For the MFs of a Gaussian map containing a disk, we propose the decomposition

$$\bar{v}_i(\nu) := \left(1 - \frac{A}{4\pi}\right) \bar{v}_i^G(\nu) + \frac{A}{4\pi} \bar{v}_i^G(\nu - \delta T) + \partial A_i(\nu, \delta T, \text{shape}), \quad i \in \{0, 1, 2\}.$$

The first two terms are area weighted MFs of pure Gaussian fields. The first term represents the part of the sky which is unaffected by the collision. The second term corresponds to the MFs of a Gaussian field in A with the mean temperature shifted by δT.

The third term ∂A_i stands for the "boundary" effects of the transition region where the temperature drops from $\delta T \to 0$. Note that, within this decomposition, information about the shape of the collision region is entirely contained in ∂A_i since the first two terms solely depend on the constant temperature shift δT and the area of the collision region $A = 2\pi (1 - \cos \vartheta_D)$ for a disk with opening angle ϑ_D.

To numerically generate a map that contains a disk we linearly add a signal u_D into a map of gaussian temperature anisotropies u_G with

$$u_D(\vartheta) := \delta T \cdot \Theta(\vartheta_D - \vartheta).$$

The signal defined in this way is similar to the form proposed in equation (3.2). In practice this is done in spherical harmonic (a_{lm}) space via the sum $a_{lm} = a_{lm}^G + a_{lm}^D$, where a_{lm}^G is the gaussian spectrum and

$$a_{lm}^D = \delta T \sqrt{\frac{\pi}{2l+1}} \left(P_{l-1}(\cos \vartheta_D) - P_{l+1}(\cos \vartheta_D)\right) \delta_{m0},$$

is the disk spectrum. Both spectra are cut off at the some high $l_{\max} > 1000$, which has to be high enough to ensure that the steepness of the step function is preserved. While in principle, there exist a small contribution from the boundary region of the disk, as the transition region is very small due to the steepness of the step, the signal associated with the gradient ∂A_i is highly suppressed and hence we neglect it from now on. Thus we will henceforth consider

$$\bar{v}_i(\nu, \mu, \sigma, \tau) := \left(1 - \frac{A}{4\pi}\right) \bar{v}_i^G(\nu, \mu_G, \sigma_G, \tau_G) + \frac{A}{4\pi} \bar{v}_i^G(\nu - \delta T, \mu_G, \sigma_G, \tau_G), \qquad (12.6)$$

as a sufficiently accurate approximation to the expectation values of MFs from a collision map. Recall that mean and temperature of an underlying Gaussian μ_G, σ_G are related to the mean and variance of a Gaussian field with a superimposed disk μ, σ, by

$$\mu_G = \mu - \frac{A}{4\pi} \delta T, \qquad \sigma_G = \sigma - \frac{A}{4\pi}\left(1 - \frac{A}{4\pi}\right) \delta T^2,$$

12.2. Analysis of collision maps

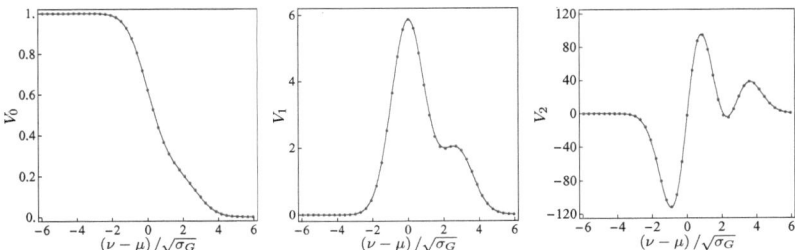

Figure 12.7.: Average numerical MFs of a Gaussian random field with a superimposed disk of $\vartheta_D = 60°$, $\delta T = 3\sqrt{\sigma_G}$ at $N_{\text{side}} = 512$ (red dots), compared to $\bar{v}_i$ (blue line) as given in equation (12.6).

and that we set $\tau_G = \tau$. The downside of equation (12.6) is that it implies that we cannot access information about the *shape* of the boundary because it is contained in the boundary terms[1]. Numerically extracted MFs for a disk of temperature difference $\delta T = 3\sqrt{\sigma_G}$ and opening angle $\vartheta_D = 60°$ are shown in Figure 12.7.

Notice that equation (12.6) is invariant under the simultaneous replacement of $\delta T \to \tilde{\delta T} = -\delta T$ and $A \to \tilde{A} = 4\pi - A$. This is a simple reflection of the fact that a "hot" spot of temperatue δT and size A in a Gaussian field with expected temperature μ_G may equally well be regarded as a "cold" spot of temperature $-\delta T$ and size $4\pi - A$ within a Gaussian field of mean temperature $\mu_G + \delta T$. This degeneracy is circumvented by restricting the consideration to disk sizes $A \leq 2\pi$, keeping in mind the implication.

12.2.2. Relative amplitude of the disk signal

In this subsection we investigate which kind of disks one can hope to detect with the help of MF statistics. Therefore it turns out to be useful to compare the MFs expected from a Gaussian field with those expected in the presence of a disk with temperature difference δT and opening angle ϑ_D. That is to consider the difference

$$\Delta v_i(\nu, \mu, \sigma, \tau, \delta T, A) := \bar{v}_i(\nu, \mu, \sigma, \tau) - \bar{v}_i^G(\nu, \mu, \sigma, \tau),$$

where μ, σ represent mean and variance of the field and τ is the variance in the gradient field of a given temperature map which is supposed to contain a disk, as calculated by equation (12.2). The disk parameters are constrained by $\sigma_G = \sigma - A/(4\pi)(1 - A/(4\pi))\delta T^2 > 0$. In Figure 12.8 we have shown the differences Δv_i for a disk with $\delta T/\sqrt{\sigma} = 0.7$ and $\vartheta_D = 50°$. Unfortunately, the shape of the difference is very similar

---
[1] The smallness of the transition region is a result of the fact that we have chosen to work with *disks* with smooth boundaries instead of some more complicated shapes – for example if the cold/hot spot is bounded by a highly irregular border with very small structures, the boundary term may contribute a non-negligible signal to the total MF. However, such shapes are not expected from generic bubble collisions scenarios.

12. Minkowski Functional statistics of a collision signal

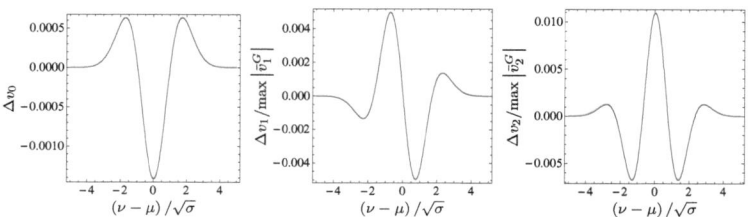

Figure 12.8.: The Figure shows the normalized difference Δv_i for $\delta T/\sqrt{\sigma} = 0.7$ and $\vartheta_D = 50°$ with the μ, σ and τ taken from the realizations presented in Figure 12.5. Notice that Δv_i is very similar in shape as $\partial_\nu^3 \bar{v}_i^G$ so that the difference that occurs in Δ_i^G may potentially be mistaken as the presence of a disk of this amplitude.

to $\partial_\nu^3 \bar{v}_i^G$ for a wide range range in the $\{\delta T, \vartheta_D\}$-parameter space. This means that the difference Δ_i^G that occurs in the numerical MFs of a Gaussian field can potentially be mistaken as the presence of a disk. Only when the temperature difference clearly obeys $\delta T/\sqrt{\sigma} \gtrsim 1$ and the disk covers a significant fraction of the sky a distinctive signature manifests itself in MFs. This has important consequences. Firstly, the temperature difference and size of a disk must be large to be detectable with MFs. Moreover, the intrinsic variation in a single outcome of numerically generated Gaussian map at WMAP resolution is large enough to mimic the presence of a prominent hot or cold spot in the MFs of the map. In other words,*MFs are not a very sensitive tool when it comes to the detection of disks in the CMB*.

12.2.3. Null test of Gaussian maps

In this subsection we will elucidate the aforementioned issues. Figures 12.9 and 12.10 show the L^2-norm of the difference

$$\Delta_i(\nu) := V_i(\nu) - \left[\bar{v}_i(\nu, \mu, \sigma, \tau) + R_i^{\Delta\nu}(\nu, \mu, \sigma, \tau)\right], \quad (12.7)$$

$$R_i^{\Delta\nu}(\nu, \mu, \sigma, \tau) := \left(1 - \frac{A}{4\pi}\right) R_i^G(\nu, \mu_G, \sigma_G, \tau_G) + \frac{A}{4\pi} R_i^G(\nu - \delta T, \mu_G, \sigma_G, \tau_G),$$

that is

$$L_i^2 := \frac{n_{bins}^{-1} \sum_j (\Delta_i(\nu_j))^2}{(\max |\bar{v}_i|)^2},$$

for single Gaussian realizations at WMAP and for an average over 1000 realizations. Figure FIG. 12.11 shows $\ln L_i^2$ for prospective PLANCK resolution ($N_{\text{side}} = 2048$).

The minimum is indicated by the green dot and therefore means a best fit of equation (12.7) to the data. The apparent presence of a disk in the case of single realizations the fluctuations in the MFs of the Gaussian field, as shown in Figure 12.5, is due to the fact that their shape is very similar to the shape in the difference Δv_i, Figure 12.8 so that

12.2. Analysis of collision maps

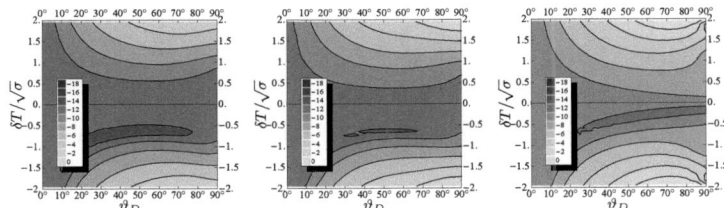

Figure 12.9.: The figure shows $\ln L_i^2$ for a single Gaussian realization and its minimum (red dot) at WMAP resolution ($N_{\text{side}} = 512$, without smoothing). The $\delta T = 0$ line is degenerate in ϑ_D space.

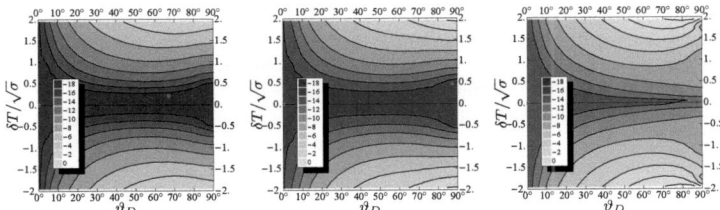

Figure 12.10.: The figure shows $\ln L_i^2$ for an average over 1000 Gaussian realizations and its minimum (red dot) at $N_{\text{side}} = 512$ without smoothing. The $\delta T = 0$ line is degenerate in ϑ_D space.

12. Minkowski Functional statistics of a collision signal

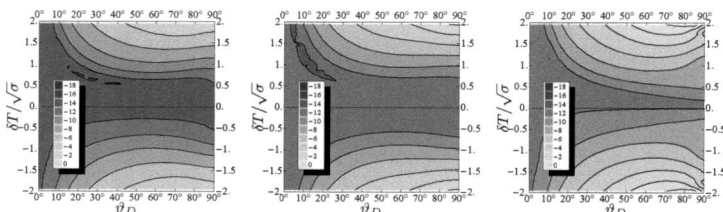

Figure 12.11.: The figure shows $\ln L_i^2$ for a single realization and its minimum (red dot) at $N_{\text{side}} = 2048$ without smoothing. The $\delta T = 0$ line is degenerate in ϑ_D space.

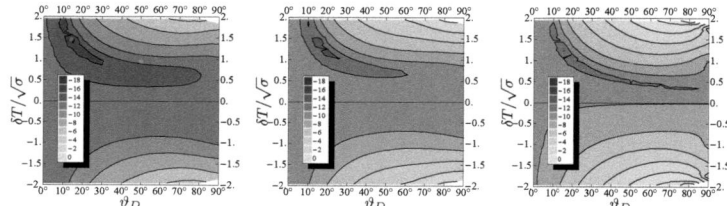

Figure 12.12.: The figure shows $\ln L_i^2$ for a single realization with $\delta T = \sqrt{\sigma}$ and $\vartheta_D = 50°$ (green dot) at $N_{\text{side}} = 512$ without smoothing. The corresponding best fit value is indicated by a red dot.

eqn. (12.6) allows for a good fit for the data. Figure 12.10 shows that upon averaging over a larger number of samples the fluctuations in the MFs decrease, cf. Figure 12.6, and the best fit essentially resembles the null result.

In Figure 12.12 and Figure 12.13 we show an example of fake collision data for $\delta T/\sqrt{\sigma} = 1$ and $\vartheta_D = 50°$. As in the Gaussian case, the remaining difference in the MFs has a severe effect on Δ_i so that the minimum of its L^2-norm (red dot) is not to be found at the input values (green dot). However, when we average the MFs taken from many realizations, the remaining difference in the MFs decreases and the minimum of the L^2-norm of Δ_i is very close to the actual input parameters. We conclude that we cannot detect disks with $\delta T/\sqrt{\sigma} \lesssim 1$, even if they cover a large fraction of the sky. Only large disks with $\vartheta_D = \mathcal{O}(10°)$ with temperature difference $\delta T/\sqrt{\sigma} \gtrsim 2$ for which the main contribution to the MFs lies clearly outside of the Gaussian, cf. Figure 12.7, can be detected with certainty. The main drawback to the use of this MFs algorithm is the remaining difference Δ_i which results in a bad signal to noise ratio for these disks. As this difference depends only weakly on the resolution we do not expect a significant improvement from PLANCK data.

12.2. Analysis of collision maps

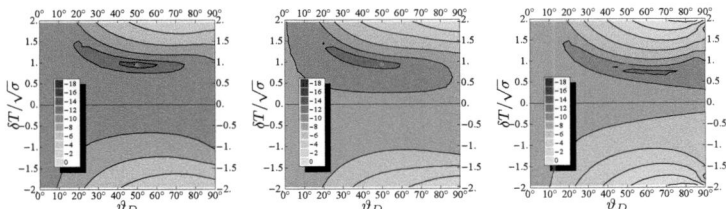

Figure 12.13.: The figure shows $\ln L_i^2$ for an average over 2048 realizations with $\delta T = \sqrt{\sigma}$ and $\vartheta_D = 50°$ (green dot) at $N_{\text{side}} = 512$ without smoothing. The corresponding best fit value is indicated by a red dot.

12.2.4. Application to WMAP seven year data

In the recent literature on the CMB cold spot, see e.g. Cruz et al. (2005, 2006, 2007), it has been argued whether the occurrence of such a spot is a likely feature of a Gaussian random field and therefore is regarded "generic", or whether it is a non-Gaussian feature the origin of which is related to a thus far unknown physical mechanism.

In this context we ask what can be inferred about the presence of spots from the MF statistics. Therefore we have used the same map making procedure as Bennett et al. (2003), Vielva et al. (2004) and Zhang and Huterer (2010) in their analysis of the CMB cold spot. We compute the temperature u of the fiducial map at x_i by the sum

$$u(x_i) = \frac{\sum_r u_r(i) w_r(i)}{\sum_r w_r(i)}.$$

of each individual differential assembly $r \in \{Q1,Q2,V1,V2,W1,W2,W3,W4\}$ weighted by $w_r(i) = N_r(i)/\sigma_r^2$, where $N_r(i)$ is the number of effective observations at pixel number i and σ_r is the noise dispersion of the respective receiver. The maps are added at a resolution of $N_{\text{side}} = 512$ and smoothed with $\vartheta_s = 1°$. The a_{lm}'s are extracted before the KQ75 mask is applied. The MFs are then computed by summing only over the unmasked pixels, i.e.

$$V_i(\nu) := \frac{\sum_{j=1}^{N_{\text{pix}}} W_j \mathcal{I}_i(\nu, x_j)}{\sum_{j=1}^{N_{\text{pix}}} W_j},$$

with $W_j = 1$ when the pixel is not hidden behind the mask and 0 otherwise.

It is clear that MFs do not have the sensitivity to pick up the small signal as seen by Vielva et al. (2004), i.e. a $\delta T = -0.016$mK at 5 degrees at roughly 3σ, since in this regime the signal is smaller than remaining noise of a single realization as described in section (12.2.2). Indeed, fitting the co-added map into our estimator, we obtained a disk with temperature difference $\delta T \simeq -0.063$mK and opening angle $\vartheta_D \simeq 35°$, which is clearly a fit to noise and hence is not physical.

13. Summary and conclusion of Part II

Motivated by recent work on cosmic bubble collisions and their potentially observable signatures in the CMB, we studied the utility of MFs for their detection. Therefore we have given a short introduction to MFs, the HEALPix software package and the algorithm we have used to compute MFs numerically from maps in the HEALPix representation. For illustration, we have applied our code on earth's topographical map.

Further on, we have presented the expectation values for Minkowski Functionals of Gaussian random fields on the two-sphere. We then resolved the long-standing issue with the MF "residuals" – systematic differences between analytically and numerically computed MF which are independent of map resolution and sample sizes. We show that these residuals are in fact a result of finite bin-sizes, and not caused by pixelation, masking or other intangible effects as originally suspected. We derive a *map-independent* analytical formula to characterize these residuals at all orders, allowing one to convolve the effects of bin-size into the MF estimators.

After removal of these residuals, we find that the remaining descrepancies between the analytic estimates and the numerical MF are of order $\mathcal{O}\left(10^{-3}\right)$. This descrepancies is proportional to the number of pixels of the map and the number of sample sizes, indicating that we are approaching the limit expected from random noise alone. Unfortunately, as we demonstrated in the text, this noise has a characteristic that is roughly similar to the expected disk signal, and hence severely limits our ability to probe small disk signals.

We apply our residual-free MF estimator to the investigation of Gaussian temperature fluctuations containing a superimposed collision signal. To characterize the signal-to-noise of our estimators, we generated collision maps by modeling the signal as a uniform shift of mean temperature within a circular spot (a disk) in an otherwise Gaussian field. We find that our least-squares fitting procedure accurately reproduces the underlying signal only when a large number of realizations of maps are averaged over. For a single WMAP and PLANCK resolution map we are able to recover the result only when a highly prominent disk, with $|\delta T| \gtrsim 2\sqrt{\sigma_G}$ and $\vartheta_D \gtrsim 40°$ is present. This is unfortunate, as it means that MF are intrinsically too noisy to be able to distinguish cold and hot spots in the CMB for small sizes as suggested by Vielva et al. (2004). In order to confirm our suspicion, we apply our prescription to WMAP7 map and find that we do not recover the latter's conclusions.

Appendix

A. Spherically symmetric spacetime

Under a coordinate transformation f from a manifold onto itself, the components of the metric tensor change as
$$g_{\mu\nu}(p) = \tilde{g}_{\alpha\beta}(f(p))\frac{\partial y^\alpha}{\partial x^\mu}\frac{\partial y^\beta}{\partial x^\nu},$$
where x and y are the coordinates of p and $f(p)$ respectively. If $\tilde{g}_{\alpha\beta}(f(p)) = g_{\alpha\beta}(f(p))$ the coordinate transformation is called an isometry. Accordingly, the infinitesimal coordinate transformation $x^\mu \mapsto x^\mu + \epsilon\xi^\mu$ is an isometry if
$$\xi^\alpha \partial_\alpha g_{\mu\nu} + g_{\alpha\nu}\partial_\mu \xi^\alpha + g_{\alpha\mu}\partial_\nu \xi^\alpha = 0,$$
or, in an explicitly covariant form
$$\nabla_\mu \xi_\nu + \nabla_\nu \xi_\mu = 0.$$
This is Killing's equation and a vector field satisfying it is called a Killing vector field. It means that the geometry does not change as one moves along the flow lines of ξ^μ. In this sense, a Killing vector field represents the direction of a symmetry in a spacetime.

The isometries generated by the Killing vector fields form a group, the group of motions of the spacetime. Since the Killing equation is linear and homogeneous, linear combinations of Killing vector fields are again Killing vector fields and the general solution can be expressed as a linear combination of basis solutions. In addition, the commutator of two Killing vector fields is also a Killing vector field: $[\boldsymbol{\xi_A}, \boldsymbol{\xi_B}] = C_{AB}^D \boldsymbol{\xi_D}$. The constants C_{AB}^D are the structure constants characterizing the group of motions. They are independant of the choice of coordinates, but do depend upon the choice of basis of Killing vector fields.

In an n-dimensional spacetime there may exist up to $n(n + 1)/2$ Killing vector fields. Those spacetimes which admit the maximum number of Killing vectors are called maximally symmetric spacetimes. The Killing equation can be used to solve two kinds of problems: Derive the symmetries of a given metric, or derive a metric with given symmetries.

The general line element for spherically symmetric spacetimes follows from the solution of Killings equation with Killing vector fields obeying the Lie algebra of the rotation group $O(3)$. The structure constants equal the Levi-Cevita symbol $C_{AB}^D = \epsilon_{ABD}$ and in spherical coordinates the Killing vector fields can be written as
$$\begin{aligned}\mathbf{L_x} &= -\cos\varphi\,\boldsymbol{\partial}_\vartheta + \sin\varphi\cot\vartheta\,\boldsymbol{\partial}_\varphi,\\ \mathbf{L_y} &= \sin\varphi\,\boldsymbol{\partial}_\vartheta + \cos\varphi\cot\vartheta\,\boldsymbol{\partial}_\varphi,\\ \mathbf{L_z} &= \boldsymbol{\partial}_\varphi.\end{aligned}$$

A. Spherically symmetric spacetime

Killing's equations for a metric $g_{\mu\nu}(t, r, \vartheta, \varphi)$ imply

$$-\partial_\vartheta g_{\mu\nu} + \cot\vartheta \left(g_{\varphi\nu}\delta_\mu^\varphi + g_{\varphi\mu}\delta_\nu^\varphi \right) = 0. \tag{A.2a}$$

$$g_{\vartheta\nu}\delta_\mu^\varphi + g_{\vartheta\mu}\delta_\nu^\varphi - \frac{1}{\sin^2\vartheta} \left(g_{\varphi\nu}\delta_\mu^\vartheta + g_{\varphi\mu}\delta_\nu^\vartheta \right) = 0. \tag{A.2b}$$

By inspection of equation (A.2a) for $\mu \neq \varphi \neq \nu$ it follows that $\partial_\vartheta g_{\mu\nu} = 0$. Whereas from equation (A.2b) we have

- for $\mu = \vartheta$ and $\nu \neq \vartheta, \varphi \Rightarrow g_{\varphi\nu} = g_{\nu\varphi} = 0$,
- for $\mu = \varphi$ and $\nu \neq \vartheta, \varphi \Rightarrow g_{\vartheta\nu} = g_{\nu\vartheta} = 0$,
- for $\mu = \vartheta = \nu \Rightarrow g_{\vartheta\varphi} = g_{\varphi\vartheta} = 0$,
- for $\mu = \varphi = \nu \Rightarrow g_{\varphi\varphi} = \sin^2\vartheta\, g_{\vartheta\vartheta}$.

Hence, the general spherically symmetric line element may be written as

$$ds^2 = -N^2 dt^2 + 2\beta dt dr + \alpha^2 dr^2 + R^2 d\Omega^2, \tag{A.3}$$

The metric functions N, α, β, R depend solely on t and r. The Killing vector fields act on the $t =$ constant, $r =$ constant hypersurfaces. The centers of symmetry are where $R = 0$, but such points must not necessarily exist.

Additionally, we can switch to a comoving, synchronos frame by a coordinate transformation in the (t, r) subspace if there exists a rotationless, timelike vector field, like the velocity field of a perfect fluid. This eliminates the off-diagonal term in the metric

$$ds^2 = -N^2 dt^2 + \alpha^2 dr^2 + R^2 d\Omega^2. \tag{A.4}$$

Given that R is not a constant one could further simplify this metric. In regions where the gradient of R is timelike R may be used as the new time coordinate, and in regions where the gradient of R is spacelike it can be used as the new radial coordinate.

B. Derivation of the Lemaître-Tolman spacetime

Starting from the comoving line element of a spherically symmetric spacetime with a dust source (A.4) we note that the property $u^\mu \nabla_\mu u^\nu = 0$ implies $\partial_r N = 0$. Thus, with rescaling of the time coordinate we can set $N = 1$ without loss of generality. A careful look at the tr-component of Einstein's equations reveals the implication

$$\partial_t \left(\alpha^{-1} \partial_r R \right) = 0,$$

and thus one may write

$$\alpha = \frac{\partial_r R}{\sqrt{1 + 2E}},$$

where $E(r)$ depends on the radial coordinate only. This establishes the line element for the Lemaître-Tolman spacetime

$$ds^2 = -dt^2 + \frac{(\partial_r R(t,r))^2}{1 + 2E(r)} dr^2 + R^2(t,r) d\Omega^2.$$

For completenes we will write down the relevant geometrical quantities for this spacetime. Using the Levi-Civitá connection in a coordinate basis the Christoffel symbols follow from

$$\Gamma^\lambda_{\mu\nu} = \frac{1}{2} g^{\lambda\alpha} \left(\partial_\mu g_{\nu\alpha} + \partial_\nu g_{\mu\alpha} - \partial_\alpha g_{\mu\nu} \right).$$

Those that do not vanish are

$$\Gamma^t_{rr} = \frac{R' \dot{R}'}{1 + 2E}, \quad \Gamma^t_{\vartheta\vartheta} = R\dot{R}, \quad \Gamma^t_{\varphi\varphi} = \sin^2 \vartheta \, \Gamma^t_{\vartheta\vartheta},$$

$$\Gamma^r_{tr} = \Gamma^r_{rt} = \frac{\dot{R}'}{R'}, \quad \Gamma^r_{rr} = \frac{R''}{R'} - \frac{E'}{1 + 2E}, \quad \Gamma^r_{\vartheta\vartheta} = -\frac{R}{R'}(1 + 2E),$$

$$\Gamma^r_{\varphi\varphi} = \sin^2 \vartheta \, \Gamma^r_{\vartheta\vartheta}, \quad \Gamma^\vartheta_{\varphi\varphi} = -\cos\vartheta \sin\vartheta, \quad \Gamma^\varphi_{\vartheta\varphi} = \Gamma^\varphi_{\varphi\vartheta} = \cot\vartheta,$$

$$\Gamma^\vartheta_{t\vartheta} = \Gamma^\vartheta_{\vartheta t} = \Gamma^\varphi_{t\varphi} = \Gamma^\varphi_{\varphi t} = \frac{\dot{R}}{R}, \quad \Gamma^\vartheta_{r\vartheta} = \Gamma^\vartheta_{\vartheta r} = \Gamma^\varphi_{r\varphi} = \Gamma^\varphi_{\varphi r} = \frac{R'}{R}.$$

An overdot and prime represent a partial derivative with respect to t and r respectively. We will skip writing down the Riemann tensor and give the Ricci tensor $R_{\mu\nu} := R^\alpha_{\beta\alpha\nu}$

$$R_{tt} = -\frac{2\ddot{R}}{R} - \frac{2\ddot{R}'}{R'}, \quad R_{rr} = \frac{R'\left(R\ddot{R}' + 2\dot{R}\dot{R}' - 2E'\right)}{R(1 + 2E)},$$

$$R_{\vartheta\vartheta} = \dot{R}^2 + R\left(\ddot{R} + \frac{\dot{R}\dot{R}' - E'}{R'}\right) - 2E, \quad R_{\varphi\varphi} = \sin^2 \vartheta \, R_{\vartheta\vartheta}.$$

B. Lemaître-Tolman spacetime

and the curvature scalar $\mathcal{R} := g^{\alpha\beta} R_{\alpha\beta}$.

$$\mathcal{R} = \frac{2}{R}\left(2\ddot{R} + \frac{\dot{R}^2 - 2E}{R} + \frac{2\dot{R}\dot{R}' + R\ddot{R}' - 2E'}{R'}\right).$$

Collecting terms we get the non-vanishing components of the Einstein Tensor $G_{\mu\nu} := R_{\mu\nu} - g_{\mu\nu}\mathcal{R}/2$

$$G_{tt} = \frac{\dot{R}^2 - 2E}{R^2} + \frac{2\dot{R}\dot{R}' - 2E'}{RR'}, \quad G_{rr} = \frac{R'^2\left(2E - \dot{R}^2 - 2R\ddot{R}\right)}{R^2(1 + 2E)},$$

$$G_{\vartheta\vartheta} = \frac{R}{R'}\left(E' - \dot{R}\dot{R}' - R'\ddot{R} - R\ddot{R}'\right), \quad G_{\varphi\varphi} = \sin^2\vartheta\, G_{\vartheta\vartheta}.$$

Together with the Stress-energy tensor of dust in the comoving coordinates $T_{\mu\nu} = \rho \delta^t_\mu \delta^t_\nu$ the Einstein equations $G_{\mu\nu} + \Lambda g_{\mu\nu} = 8\pi T_{\mu\nu}$ reduce to

$$\frac{\dot{R}^2 - 2E}{R^2} + \frac{2\dot{R}\dot{R}' - 2E'}{RR'} - \Lambda = 8\pi\rho, \tag{B.1a}$$

$$2E - \dot{R}^2 - 2R\ddot{R} + \Lambda R^2 = 0. \tag{B.1b}$$

The latter has the first integral

$$\dot{R}^2 = 2E + \frac{2M}{R} + \frac{\Lambda}{3}R^2,$$

where M is an arbitrary function of r. Equation (B.1a) simplifies to

$$8\pi\rho = \frac{2M'}{R^2 R'}.$$

These are the governing equations of the Lemaître-Tolman spacetime.

C. Foliations of de Sitter spacetime

De Sitter spacetime can be embedded in five dimensional Minkowski spacetime as the hyperboloid $g_{\mu\nu}X^\mu X^\nu = H^{-2}$, where $H := \sqrt{\Lambda/3}$ is a constant. Due to its high symmetry de Sitter spacetime admits many convenient foliations. We will illustrate the embedding coordinates and induced metrics of the so called static, open, flat and closed slicings.

- Static slicing:
$$X^0 = H^{-1}\sqrt{1 - H^2 r^2}\sinh(Ht) ,$$
$$X^4 = H^{-1}\sqrt{1 - H^2 r^2}\cosh(Ht) ,$$
$$X^i = r\omega_i, \quad \omega_i = (\cos\varphi\sin\vartheta, \sin\varphi\sin\vartheta, \cos\vartheta) ,$$
$$ds^2 = -\left(1 - H^2 r^2\right)dt^2 + \left(1 - H^2 r^2\right)^{-1}dr^2 + r^2 d\Omega^2 .$$

- Open slicing:
$$X^0 = H^{-1}\sinh(Ht)\cosh(\chi) ,$$
$$X^4 = H^{-1}\cosh(Ht) ,$$
$$X^i = H^{-1}\sinh(Ht)\sinh(\chi)\omega_i ,$$
$$ds^2 = -dt^2 + H^{-2}\sinh^2(Ht)\left(d\chi^2 + \sinh^2(\chi) d\Omega^2\right) .$$

- Flat slicing:
$$X^0 = H^{-1}\sinh(Ht) + Hr^2 \exp(Ht)/2 ,$$
$$X^4 = H^{-1}\cosh(Ht) - Hr^2 \exp(Ht)/2 ,$$
$$X^i = r\exp(Ht)\omega_i ,$$
$$ds^2 = -dt^2 + \exp(2Ht)\left(dr^2 + r^2 d\Omega^2\right) .$$

- Closed slicing:
$$X^0 = H^{-1}\sinh(Ht) ,$$
$$X^4 = H^{-1}\cosh(Ht)\cos(\psi) ,$$
$$X^i = H^{-1}\sinh(Ht)\sin(\psi)\omega_i ,$$
$$ds^2 = -dt^2 + H^{-2}\cosh^2(Ht)\left(d\psi^2 + \sin^2(\psi) d\Omega^2\right) .$$

The open, flat and closed slicings have Robertson-Walker symmetry with $k < 0$, $k = 0$ and $k > 0$. The closed slicing is geodesically complete and thus covers the whole spacetime, whereas the other slicings cover only a part of the spacetime as can be seen in Figure C.1.1.

C. Foliations of de Sitter spacetime

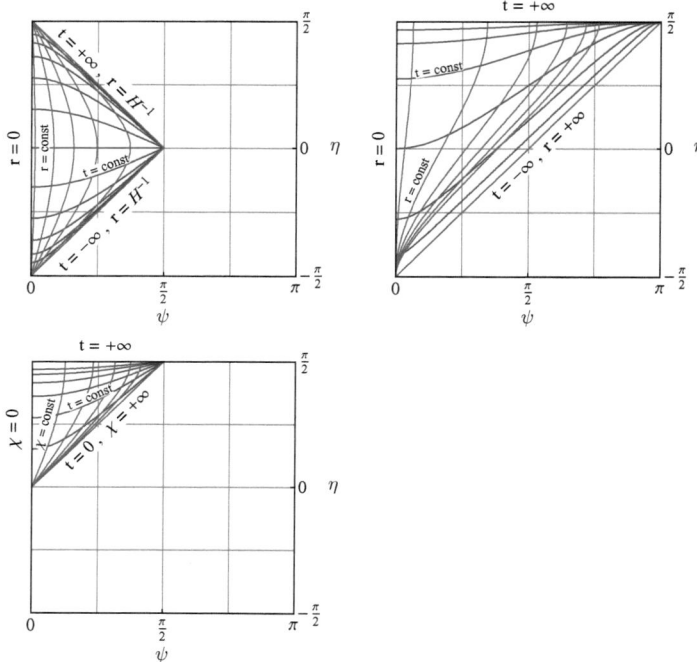

Figure C.1.: Illustration of the different slicings of de Sitter spacetime. The closed foliation covers the entire spacetime, whereas the other slicings cover only part of it. It is shown how the coordinates of these slicings relate to those of the conformal closed slicing given by ψ and $\eta = \int dt/a$. Upper left: Static slicing. Upper right: Flat slicing. Lower: Open slicing.

D. General Relativistic Junction Conditions

Let $\left(\mathcal{M}^{\pm}, g^{\pm}_{\mu\nu}, x^{\mu}_{\pm}\right)$ represent two spacetimes that are valid solutions to Einstein's field equations with metric $g^{\pm}_{\mu\nu}$ and coordinates $x^{\mu}_{\pm}$. Define a hypersurface Σ that separates each spacetime into two parts such that

$$\mathcal{M}^{\pm} = \mathcal{M}^{\pm}_1 \cup \mathcal{M}^{\pm}_2 \quad \text{and} \quad \Sigma^{\pm} = \partial\mathcal{M}^{\pm}_1 \cap \partial\mathcal{M}^{\pm}_2\,.$$

Now, one may ask: What are the conditions under which a spacetime given by some union of $\mathcal{M}^{-}_{1/2} \cup \mathcal{M}^{+}_{1/2}$ with the identification $\Sigma^{+} = \Sigma^{-} := \Sigma$ is a valid solution to the field equations. This task is solved and well established since the work of Israel (1966). We will give a brief introduction to the subject.

Let $\left(\mathcal{M}^{\pm}, g^{\pm}_{\mu\nu}, x^{\mu}_{\pm}\right)$ be the parts of two spacetimes that we want to join along a hypersurface Σ. Assume that we can introduce continuous coordinates x^{μ} in a region around Σ as well as coordinates y^a on Σ. Imagine a congruence of geodesics which intersect the hypersurface orthogonally. The geodesics define a scalar field $l(x^{\mu})$ that assigns the postive (negative) proper distance from x^{μ} to Σ when $x^{\mu} \in \mathcal{M}^{+}$ ($x^{\mu} \in \mathcal{M}^{-}$). Thus, an infinitesimal displacement away from Σ can be decomposed into

$$dx^{\mu} = \frac{\partial x^{\mu}}{\partial y^a} dy^a + \frac{\partial x^{\mu}}{\partial l} dl\,.$$

We define the basis vectors along the coordinates y^a: $e^{\mu}_a := \partial x^{\mu}/\partial y^a$ and the unit normal $n^{\mu} \equiv \partial x^{\mu}/\partial l$ orthogonal to Σ. It obeys

$$n_{\alpha} n^{\alpha} := \epsilon = \begin{cases} 1, & \text{if } \Sigma \text{ is timelike}\,, \\ -1, & \text{if } \Sigma \text{ is spacelike}\,. \end{cases}$$

Junction conditions for lightlike hypersurfaces, i.e. $\epsilon = 0$, are discussed in Barrabés and Israel (1991); Poisson (2002). More details can be found in the textbook by Poisson (2004). Without loss of generality we can define n^{μ} to point from $\mathcal{M}^{-}$ to $\mathcal{M}^{+}$. By construction we have $n_{\alpha} e^{\alpha}_a = 0$. For clarity we define for any tensorial quantity T

$$[T] := T^{+}|_{\Sigma} - T^{-}|_{\Sigma} \quad \text{and} \quad \{T\} := \frac{1}{2}\left(T^{+}|_{\Sigma} + T^{-}|_{\Sigma}\right)\,.$$

The first condition for a valid junction is continuity of the induced metric on the hypersurface, which is clearly necessary if Σ is to have a well-defined geometry. Hence, the first junction condition reads

$$\left[g_{\alpha\beta} e^{\alpha}_a e^{\beta}_b\right] = [h_{ab}] = 0\,. \tag{D.1}$$

D. Relativistic Junction Conditions

Moreover, to obtain a completely smooth junction, we also require that the embedding of the hypersurface is continuous. This means that the extrinsic curvature tensors match on Σ: $[K_{ab}] = 0$. However, if this condition is violated it can be given a sound physical interpretation. In presence of a jump discontinuity in K_{ab} the hypersurface can be associated with a thin surface layer that has a stress-energy tensor given by

$$8\pi\epsilon S_{ab} := [K_{ab}] - h_{ab}[K]\,. \tag{D.2}$$

This represents the second junction condition. The components of the extrinsic curvature tensor K_{ab} are the projection of the covariant derivative of the vector e_a^α along e_b^β onto the normal n^λ

$$K_{ab} := n_\lambda \Gamma^\lambda_{\alpha\beta} e_a^\alpha e_b^\beta\,. \tag{D.3}$$

Thus, for Einstein's equations to be fulfilled on the hypersurface, the full stress-energy tensor has to be

$$T_{\mu\nu} := \Theta(l)T^+_{\mu\nu} + \Theta(-l)T^-_{\mu\nu} + \delta(l)S_{\mu\nu}\,,$$

where $S^{\mu\nu} = S^{ab}e_a^\mu e_b^\nu$. The other projections of Einstein's equations then imply

$$^{(3)}\nabla_b S_a^b + \left[T_{\alpha\beta}e_a^\alpha n^\beta\right] = 0\,, \tag{D.4a}$$

$$S_{ab}\left\{K^{ab}\right\} + \left[T_{\alpha\beta}n^\alpha n^\beta\right] = 0\,. \tag{D.4b}$$

Together these equations determine the evolution of Σ in $\mathcal{M}^\pm$.

E. Derivation of Minkowski functionals of a Gaussian and an axially symmetric field on the sphere

E.1. Transformation to surface integrals

In this section we will show how the boundary integral that appears in the second and third Minkowski functional as

$$\int_{\partial Q_\nu} dl = \int_{S^2} d\Omega\, \delta(u - \nu) |\nabla u| , \qquad \text{(E.1)}$$

can be expresssed as integrals over the sphere. We show that the identity (E.1) is true in two-dimensional Euclidean space. Assume that we have local coordinate chart $\{x, y\}$ in which $\partial_y u \neq 0$. This implicitly defines a function f through $u(x, f(x)) = \nu$ so that

$$\partial_x u(x, f(x)) + \partial_y u(x, f(x)) \frac{df(x)}{dx} = 0.$$

The right hand side of equation (E.1) can be transformed by

$$\int dx \int dy\, \delta(u - \nu) |\nabla u| = \int dx \int dy\, \frac{\delta(y - f(x))}{|\partial_y u|} \sqrt{(\partial_x u)^2 + (\partial_y u)^2} ,$$

$$= \int dx \sqrt{1 + \left(\frac{\partial_x u(x, f(x))}{\partial_y u(x, f(x))}\right)^2} ,$$

$$= \int dx \sqrt{1 + \left(\frac{df}{dx}\right)^2} = \int dl .$$

The generalization to a curved space is straightforward.

E. Minkowski functionals on the sphere

E.2. Gaussian Random fields

Let u be a Gaussian random field with vanishing mean and correlation function

$$\xi(\vartheta) := \langle u(x)u(y)\rangle = \frac{1}{4\pi}\sum_{l=1}^{\infty}(2l+1)C_l P_l(\cos\vartheta),$$

where ϑ is the geodetic distance between x and y. Following the approach of Tomita (1986) we note that the six variables $U := \{u, u_{;\vartheta}, u_{;\varphi}, u_{;\vartheta\vartheta}, u_{;\varphi\varphi}, u_{;\vartheta\varphi}\}$ are jointly Gaussian distributed. The respective covariance matrix is

$$\sigma_{ij} = \begin{pmatrix} \sigma & 0 & 0 & -\tau & -\tau & 0 \\ 0 & \tau & 0 & 0 & 0 & 0 \\ 0 & 0 & \tau & 0 & 0 & 0 \\ -\tau & 0 & 0 & v & v/3 & 0 \\ -\tau & 0 & 0 & v/3 & v & 0 \\ 0 & 0 & 0 & 0 & 0 & v/3 \end{pmatrix},$$

with $\sigma := \xi(0), \tau := |\xi''(0)|$ and $v := \xi^{(4)}(0)$ are given by the correlation function. The average Minkowski functionals of the field u, cf. equations (10.2) – (10.4), can be calculated with the help of the probability distribution of U

$$P(U) = \frac{1}{\sqrt{(2\pi)^6 \det \sigma_{ij}}}\exp\left(-\frac{1}{2}\sigma_{ij}^{-1}U_i U_j\right),$$

by

$$\left\langle v_0^G(\nu)\right\rangle = \int dU\, P(U)\Theta(u-\nu),$$

$$\left\langle v_1^G(\nu)\right\rangle = \int dU\, P(U)\delta(u-\nu)\,|\nabla u|,$$

$$\left\langle v_2^G(\nu)\right\rangle = \int dU\, P(U)\delta(u-\nu)\,|\nabla u|\,\kappa.$$

Explicitly, the norm of the gradient of u is $|\nabla u| = \sqrt{u_{;\vartheta}^2 + u_{;\varphi}^2}$ and the geodesic curvature is given by equation (10.8). The integrals yield

$$\left\langle v_0^G(\nu)\right\rangle = \frac{1}{2}\left(1 - \mathrm{erf}\left(\frac{\nu}{\sqrt{2\sigma}}\right)\right),$$

$$\left\langle v_1^G(\nu)\right\rangle = \frac{1}{8}\sqrt{\frac{\tau}{\sigma}}\exp\left(-\frac{\nu^2}{2\sigma}\right),$$

$$\left\langle v_2^G(\nu)\right\rangle = \frac{\tau}{2\pi^{3/2}\sigma}\frac{\nu}{\sqrt{2\sigma}}\exp\left(-\frac{\nu^2}{2\sigma}\right),$$

where erf is the Gaussian error function $\mathrm{erf}(x) = \frac{2}{\sqrt{\pi}}\int_0^x dt\,\exp(-t^2)$.

Bibliography

Abbott, L. F., Harari, D., and Park, Q. (1987). Vacuum decay in curved backgrounds. *Classical and Quantum Gravity*, 4:L201–L204.

Aguirre, A. and Johnson, M. C. (2005). Dynamics and instability of false vacuum bubbles. *Phys. Rev. D*, 72(10):103525–+.

Aguirre, A. and Johnson, M. C. (2006). Two tunnels to inflation. *Phys. Rev. D*, 73(12):123529–+.

Aguirre, A. and Johnson, M. C. (2008). Towards observable signatures of other bubble universes. II. Exact solutions for thin-wall bubble collisions. *Phys. Rev. D*, 77(12):123536–+.

Aguirre, A. and Johnson, M. C. (2009). A status report on the observability of cosmic bubble collisions. *arxiv/hep-th:0908.4105*.

Aguirre, A., Johnson, M. C., and Shomer, A. (2007). Towards observable signatures of other bubble universes. *Phys. Rev. D*, 76(6):063509–+.

Aguirre, A., Johnson, M. C., and Tysanner, M. (2009). Surviving the crash: Assessing the aftermath of cosmic bubble collisions. *Phys. Rev. D*, 79(12):123514–+.

Albrecht, A. and Steinhardt, P. J. (1982). Cosmology for grand unified theories with radiatively induced symmetry breaking. *Physical Review Letters*, 48:1220–1223.

Aurilia, A., Palmer, M., and Spallucci, E. (1989). Evolution of bubbles in a vacuum. *Phys. Rev. D*, 40:2511–2518.

Bardeen, J. M., Steinhardt, P. J., and Turner, M. S. (1983). Spontaneous creation of almost scale-free density perturbations in an inflationary universe. *Phys. Rev. D*, 28:679–693.

Barrabés, C. and Israel, W. (1991). Thin shells in general relativity and cosmology: The lightlike limit. *Phys. Rev. D*, 43:1129–1142.

Barrow, J. D. and Stein-Schabes, J. (1984). Inhomogeneous cosmologies with cosmological constant. *Physics Letters A*, 103:315–317.

Bassett, B. A., Tsujikawa, S., and Wands, D. (2006). Inflation dynamics and reheating. *Reviews of Modern Physics*, 78:537–589.

Baumann, D. (2009). TASI Lectures on Inflation. *arxiv:hep-th/0907.5424*.

Bibliography

Bennett, C. L., Hill, R. S., Hinshaw, G., Larson, D., Smith, K. M., Dunkley, J., Gold, B., Halpern, M., Jarosik, N., Kogut, A., Komatsu, E., Limon, M., Meyer, S. S., Nolta, M. R., Odegard, N., Page, L., Spergel, D. N., Tucker, G. S., Weiland, J. L., Wollack, E., and Wright, E. L. (2010). Seven-Year Wilkinson Microwave Anisotropy Probe (WMAP) Observations: Are There Cosmic Microwave Background Anomalies? *arxiv:astro-ph.co/1001.4758*.

Bennett, C. L., Hill, R. S., Hinshaw, G., Nolta, M. R., Odegard, N., Page, L., Spergel, D. N., Weiland, J. L., Wright, E. L., Halpern, M., Jarosik, N., Kogut, A., Limon, M., Meyer, S. S., Tucker, G. S., and Wollack, E. (2003). First-Year Wilkinson Microwave Anisotropy Probe (WMAP) Observations: Foreground Emission. *Astrophysical Journal Supplement*, 148:97–117.

Berezin, V. A., Kuzmin, V. A., and Tkachev, I. I. (1987). Dynamics of bubbles in general relativity. *Phys. Rev. D*, 36:2919–2944.

Bernardeau, F. and Uzan, J. (2003). Inflationary models inducing non-Gaussian metric fluctuations. *Phys. Rev. D*, 67(12):121301–+.

Blau, S. K., Guendelman, E. I., and Guth, A. H. (1987). Dynamics of false-vacuum bubbles. *Phys. Rev. D*, 35:1747–1766.

Bondi, H. (1947). Spherically symmetrical models in general relativity. *Mon. Not. Roy. Astron. Soc.*, 107:410–425.

Borde, A., Guth, A. H., and Vilenkin, A. (2003). Inflationary Spacetimes Are Incomplete in Past Directions. *Physical Review Letters*, 90(15):151301–+.

Bousso, R. (2008). The cosmological constant. *General Relativity and Gravitation*, 40:607–637.

Bousso, R. and Polchinski, J. (2000). Quantization of four-form fluxes and dynamical neutralization of the cosmological constant. *Journal of High Energy Physics*, 6:6–+.

Brown, A. R., Sarangi, S., Shlaer, B., and Weltman, A. (2007). Enhanced Brane Tunneling and Instanton Wrinkles. *Physical Review Letters*, 99(16):161601–+.

Bucher, M., Goldhaber, A. S., and Turok, N. (1995). Open universe from inflation. *Phys. Rev. D*, 52:3314–3337.

Calzetta, E. and Sakellariadou, M. (1992). Inflation in inhomogeneous cosmology. *Phys. Rev. D*, 45:2802–2805.

Carr, B. (2008). Universe or Multiverse? *ISBN:978-052-184-841-1. 505pp. Cambridge University Press, 2008*.

Carroll, S. M. (2001). The Cosmological Constant. *Living Reviews in Relativity*, 4:1–+.

Chang, S., Kleban, M., and Levi, T. S. (2008). When Worlds Collide. *Journal of Cosmology and Astroparticle Physics*, 0804:034.

Chang, S., Kleban, M., and Levi, T. S. (2009). Watching worlds collide: effects on the CMB from cosmological bubble collisions. *Journal of Cosmology and Astroparticle Physics*, 4:25–+.

Chen, X. (2010). Primordial Non-Gaussianities from Inflation Models. *Advances in Astronomy*, 2010.

Coleman, S. (1977). Fate of the false vacuum: Semiclassical theory. *Phys. Rev. D*, 15:2929–2936.

Coleman, S. and de Luccia, F. (1980). Gravitational effects on and of vacuum decay. *Phys. Rev. D*, 21:3305–3315.

Cruz, M., Cayón, L., Martínez-González, E., Vielva, P., and Jin, J. (2007). The Non-Gaussian Cold Spot in the 3 Year Wilkinson Microwave Anisotropy Probe Data. *Astrophysical Journal*, 655:11–20.

Cruz, M., Martínez-González, E., Vielva, P., and Cayón, L. (2005). Detection of a non-Gaussian spot in WMAP. *Mon. Not. Roy. Astron. Soc.*, 356:29–40.

Cruz, M., Tucci, M., Martínez-González, E., and Vielva, P. (2006). The non-Gaussian cold spot in WilkinsonMicrowaveAnisotropyProbe: significance, morphology and foreground contribution. *Mon. Not. Roy. Astron. Soc.*, 369:57–67.

Czech, B., Kleban, M., Larjo, K., Levi, T. S., and Sigurdson, K. (2010). Polarizing Bubble Collisions. *arxiv:astro-ph.co/1006.0832*.

Durrer, R. (2008). The Cosmic Microwave Background. *ISBN:978-052-184-704-9. 424pp. Cambridge University Press, 2008.*

Easther, R., Giblin, Jr., J. T., Hui, L., and Lim, E. A. (2009). New mechanism for bubble nucleation: Classical transitions. *Phys. Rev. D*, 80(12):123519–+.

Ellis, G. F. R. (2003). A historical review of how the cosmological constant has fared in general relativity and cosmology. *Chaos Solitons and Fractals*, 16:505–512.

Ellis, G. F. R. and Madsen, M. S. (1991). Exact scalar field cosmologies. *Classical and Quantum Gravity*, 8:667–676.

Ellis, G. F. R., Nel, S. D., Maartens, R., Stoeger, W. R., and Whitman, A. P. (1985). Ideal observational cosmology. *Physics Reports*, 124:315–417.

Ellis, G. F. R. and Stoeger, W. R. (2009). A note on infinities in eternal inflation. *General Relativity and Gravitation*, 41:1475–1484.

Eriksen, H. K., Novikov, D. I., Lilje, P. B., Banday, A. J., and Górski, K. M. (2004). Testing for Non-Gaussianity in the Wilkinson Microwave Anisotropy Probe Data: Minkowski Functionals and the Length of the Skeleton. *Astrophysical Journal*, 612:64–80.

Feeney, S. M., Johnson, M. C., Mortlock, D. J., and Peiris, H. V. (2010a). First Observational Tests of Eternal Inflation. *arXiv:astro-ph.CO/1012.1995*.

Feeney, S. M., Johnson, M. C., Mortlock, D. J., and Peiris, H. V. (2010b). First Observational Tests of Eternal Inflation: Analysis Methods and WMAP 7-Year Results. *arXiv:astro-ph.CO/1012.3667*.

Fields, B. and Sarkar, S. (2006). Big-Bang nucleosynthesis (Particle Data Group mini-review). *arXiv:astro-ph/0601514*.

Fischler, W., Paban, S., Žanić, M., and Krishnan, C. (2008). Vacuum bubble in an inhomogeneous cosmology. *Journal of High Energy Physics*, 5:41.

Freivogel, B., Horowitz, G. T., and Shenker, S. (2007). Colliding with a crunching bubble. *Journal of High Energy Physics*, 5:90–+.

Freivogel, B., Kleban, M., Nicolis, A., and Sigurdson, K. (2009). Eternal inflation, bubble collisions, and the disintegration of the persistence of memory. *Journal of Cosmology and Astroparticle Physics*, 8:36–+.

Freivogel, B., Kleban, M., Rodríguez Martínez, M., and Susskind, L. (2006). Observational consequences of a landscape. *Journal of High Energy Physics*, 3:39–+.

Friedmann, A. (1924). Über die Möglichkeit einer Welt mit konstanter negativer Krümmung des Raumes. *Zeitschrift für Physik*, 21:326–332. English translation, with historical comments: *Gen. Rel. Grav.* 31:1985, 1999.

Garriga, J., Guth, A. H., and Vilenkin, A. (2007). Eternal inflation, bubble collisions, and the persistence of memory. *Phys. Rev. D*, 76(12):123512–+.

Giblin, Jr., J. T., Hui, L., Lim, E. A., and Yang, I. (2010). How to run through walls: Dynamics of bubble and soliton collisions. *Phys. Rev. D*, 82(4):045019–+.

Goldwirth, D. S. and Piran, T. (1990). Inhomogeneity and the onset of inflation. *Physical Review Letters*, 64:2852–2855.

Goldwirth, D. S. and Piran, T. (1992). Initial conditions for inflation. *Physics Reports*, 214:223–292.

Gordon, C., Hu, W., Huterer, D., and Crawford, T. (2005). Spontaneous isotropy breaking: A mechanism for CMB multipole alignments. *Phys. Rev. D*, 72(10):103002–+.

Bibliography

Górski, K. M., Hivon, E., Banday, A. J., Wandelt, B. D., Hansen, F. K., Reinecke, M., and Bartelmann, M. (2005). HEALPix: A Framework for High-Resolution Discretization and Fast Analysis of Data Distributed on the Sphere. *Astrophysical Journal*, 622:759–771.

Gott, J. R. and Statler, T. S. (1984). Constraints on the formation of bubble universes. *Physics Letters B*, 136:157–161.

Gott, III, J. R. (1982). Creation of open universes from de Sitter space. *Nature*, 295:304–306.

Guth, A. H. (1981). Inflationary universe: A possible solution to the horizon and flatness problems. *Phys. Rev. D*, 23:347–356.

Guth, A. H. and Pi, S. (1982). Fluctuations in the new inflationary universe. *Physical Review Letters*, 49:1110–1113.

Guth, A. H. and Weinberg, E. J. (1983). Could the universe have recovered from a slow first-order phase transition? *Nuclear Physics B*, 212:321–364.

Hawking, S. W. (1982). The development of irregularities in a single bubble inflationary universe. *Physics Letters B*, 115:295–297.

Hellaby, C. (1996). The nonsimultaneous nature of the Schwarzschild R=0 singularity. *Journal of Mathematical Physics*, 37:2892–2905.

Helmer, F. and Winitzki, S. (2006). Self-reproduction in k-inflation. *Phys. Rev. D*, 74(6):063528–+.

Hikage, C., Komatsu, E., and Matsubara, T. (2006). Primordial Non-Gaussianity and Analytical Formula for Minkowski Functionals of the Cosmic Microwave Background and Large-Scale Structure. *Astrophysical Journal*, 653:11–26.

Hikage, C., Koyama, K., Matsubara, T., Takahashi, T., and Yamaguchi, M. (2009). Limits on isocurvature perturbations from non-Gaussianity in WMAP temperature anisotropy. *Mon. Not. Roy. Astron. Soc.*, 398:2188–2198.

Hikage, C., Matsubara, T., Coles, P., Liguori, M., Hansen, F. K., and Matarrese, S. (2008). Limits on primordial non-Gaussianity from Minkowski Functionals of the WMAP temperature anisotropies. *Mon. Not. Roy. Astron. Soc.*, 389:1439–1446.

Hikage, C., Schmalzing, J., Buchert, T., Suto, Y., Kayo, I., Taruya, A., Vogeley, M. S., Hoyle, F., Gott, III, J. R., and Brinkmann, J. (2003). Minkowski Functionals of SDSS Galaxies I : Analysis of Excursion Sets. *Publications of the Astronomical Society of Japan*, 55:911–931.

Hubble, E. (1929). A Relation between Distance and Radial Velocity among Extra-Galactic Nebulae. *Proceedings of the National Academy of Science*, 15:168–173.

Bibliography

Israel, W. (1966). Singular hypersurfaces and thin shells in general relativity. *Nuovo Cimento B Serie*, 44:1–14.

Kachru, S., Kallosh, R., Linde, A., and Trivedi, S. P. (2003). de Sitter vacua in string theory. *Phys. Rev. D*, 68(4):046005–+.

Komatsu, E., Dunkley, J., Nolta, M. R., Bennett, C. L., Gold, B., Hinshaw, G., Jarosik, N., Larson, D., Limon, M., Page, L., Spergel, D. N., Halpern, M., Hill, R. S., Kogut, A., Meyer, S. S., Tucker, G. S., Weiland, J. L., Wollack, E., and Wright, E. L. (2009a). Five-Year Wilkinson Microwave Anisotropy Probe Observations: Cosmological Interpretation. *Astrophysical Journal Supplement*, 180:330–376.

Komatsu, E., Dunkley, J., Nolta, M. R., Bennett, C. L., Gold, B., Hinshaw, G., Jarosik, N., Larson, D., Limon, M., Page, L., Spergel, D. N., Halpern, M., Hill, R. S., Kogut, A., Meyer, S. S., Tucker, G. S., Weiland, J. L., Wollack, E., and Wright, E. L. (2009b). Five-Year Wilkinson Microwave Anisotropy Probe Observations: Cosmological Interpretation. *Astrophysical Journal Supplement*, 180:330–376.

Kottler, F. (1918). Über die physikalischen Grundlagen der Einsteinschen Gravitationstheorie. *Annalen der Physik*, 361:401–462.

Kowalski, M., Rubin, D., Aldering, G., Agostinho, R. J., Amadon, A., Amanullah, R., Balland, C., Barbary, K., Blanc, G., Challis, P. J., Conley, A., Connolly, N. V., Covarrubias, R., Dawson, K. S., Deustua, S. E., Ellis, R., Fabbro, S., Fadeyev, V., and Fan, X. (2008). Improved Cosmological Constraints from New, Old, and Combined Supernova Data Sets. *Astrophysical Journal*, 686:749–778.

Kruskal, M. D. (1960). Maximal Extension of Schwarzschild Metric. *Physical Review*, 119:1743–1745.

Larjo, K. and Levi, T. S. (2010). Bubble, bubble, flow and Hubble: large scale galaxy flow from cosmological bubble collisions. *Journal of Cosmology and Astroparticle Physics*, 8:34–+.

Lemaître, G. (1933). L'Univers en expansion. *Annales de la Societe Scietifique de Bruxelles*, 53:51. English translation, with historical comments: *Gen. Rel. Grav.* 29:641, 1997.

Lim, E. A. and Simon, D. (2011). Can we detect Hot/Cold Spots in the CMB with Minkowski Functionals? *To appear*.

Linde, A. D. (1982). A new inflationary universe scenario: A possible solution of the horizon, flatness, homogeneity, isotropy and primordial monopole problems. *Physics Letters B*, 108:389–393.

Linde, A. D. (1983). Chaotic inflation. *Physics Letters B*, 129:177–181.

Linde, A. D. (1999). Toy model for open inflation. *Phys. Rev. D*, 59(2):023503–+.

Linde, A. D. (2008). Inflationary Cosmology. In M. Lemoine, J. Martin, & P. Peter, editor, *Inflationary Cosmology*, volume 738 of *Lecture Notes in Physics, Berlin Springer Verlag*, pages 1–54.

Linde, A. D. and Mukhanov, V. (1997). Non-Gaussian isocurvature perturbations from inflation. *Phys. Rev. D*, 56:535–+.

Madsen, M. S. (1988). Scalar fields in curved spacetimes. *Classical and Quantum Gravity*, 5:627–639.

Maldacena, J. (2003). Non-gaussian features of primordial fluctuations in single field inflationary models. *Journal of High Energy Physics*, 5:13–+.

Mantz, H., Jacobs, K., and Mecke, K. (2008). Utilizing Minkowski functionals for image analysis: a marching square algorithm. *Journal of Statistical Mechanics: Theory and Experiment*, 12:15–+.

Matsubara, T. (2010). Analytic Minkowski functionals of the cosmic microwave background: Second-order non-Gaussianity with bispectrum and trispectrum. *Phys. Rev. D*, 81(8):083505–+.

Mukhanov, V. F. and Chibisov, G. V. (1981). Quantum fluctuations and a nonsingular universe. *Soviet Journal of Experimental and Theoretical Physics Letters*, 33:532–+.

NOAA (1988). Digital relief of the Surface of the Earth. National Geophysical Data Center, Data Announcement 88-MGG-02.

Novikov, D., Feldman, H. A., and Shandarin, S. F. (1999). Minkowski Functionals and Cluster Analysis for CMB Maps. *International Journal of Modern Physics D*, 8:291–306.

Padmanabhan, T. (2003). Cosmological constant-the weight of the vacuum. *Physics Reports*, 380:235–320.

Peebles, P. J. and Ratra, B. (2003). The cosmological constant and dark energy. *Reviews of Modern Physics*, 75:559–606.

Penzias, A. A. and Wilson, R. W. (1965). A Measurement of Excess Antenna Temperature at 4080 Mc/s. *Astrophysical Journal*, 142:419–421.

Percival, W. J., Cole, S., Eisenstein, D. J., Nichol, R. C., Peacock, J. A., Pope, A. C., and Szalay, A. S. (2007). Measuring the Baryon Acoustic Oscillation scale using the Sloan Digital Sky Survey and 2dF Galaxy Redshift Survey. *Mon. Not. Roy. Astron. Soc.*, 381:1053–1066.

Plebanski, J. and Krasinski, A. (2006). An Introduction to General Relativity and Cosmology. *ISBN:978-052-185-623-2. 554pp. Cambridge University Press, 2006*.

Bibliography

Poisson, E. (2002). A reformulation of the Barrabes-Israel null-shell formalism. *arXiv:gr-qc/0207101*.

Poisson, E. (2004). A Relativist's Toolkit. *ISBN:978-052-153-780-3. 252pp. Cambridge University Press, 2004*.

Raychaudhuri, A. K. and Modak, B. (1988). Cosmological inflation with arbitrary initial conditions. *Classical and Quantum Gravity*, 5:225–232.

Robertson, H. P. (1929). On the Foundations of Relativistic Cosmology. *Proceedings of the National Academy of Science*, 15:822–829.

Rothman, T. and Ellis, G. F. R. (1986). Can inflation occur in anisotropic cosmologies? *Physics Letters B*, 180:19–24.

Saffin, P. M., Padilla, A., and Copeland, E. J. (2008). Transmission of an inhomogeneous state via resonant tunnelling. *Journal of High Energy Physics*, 9:55–+.

Schmalzing, J. and Górski, K. M. (1998). Minkowski functionals used in the morphological analysis of cosmic microwave background anisotropy maps. *Mon. Not. Roy. Astron. Soc.*, 297:355–365.

Schmalzing, J., Kerscher, M., and Buchert, T. (1996a). Minkowski Functionals in Cosmology. In S. Bonometto, J. R. Primack, & A. Provenzale, editor, *Dark Matter in the Universe*, pages 281–+.

Schmalzing, J., Kerscher, M., and Buchert, T. (1996b). Minkowski Functionals in Cosmology. In S. Bonometto, J. R. Primack, & A. Provenzale, editor, *Dark Matter in the Universe*, pages 281–+.

Schwarzschild, K. (1916). On the Gravitational Field of a Mass Point According to Einstein's Theory. *Abh. Königl. Preuss. Akad. Wissenschaften Berlin*, pages 189–196.

Starobinsky, A. A. (1980). A new type of isotropic cosmological models without singularity. *Physics Letters B*, 91:99–102.

Starobinsky, A. A. (1982). Dynamics of phase transition in the new inflationary universe scenario and generation of perturbations. *Physics Letters B*, 117:175–178.

Stoeger, W. R., Maartens, R., and Ellis, G. F. R. (1995). Proving almost-homogeneity of the universe: an almost Ehlers-Geren-Sachs theorem. *Astrophysical Journal*, 443:1–5.

Susskind, L. (2003). The Anthropic Landscape of String Theory. In *The Davis Meeting On Cosmic Inflation*.

Szekeres, G. (1960). On the singularities of a Riemannian manifold. *Publicationes Mathematicae Debrecen*, 7:285.

Tolman, R. C. (1934). Effect of Inhomogeneity on Cosmological Models. *Proceedings of the National Academy of Science*, 20:169–176. Reprint, with historical comments: *Gen. Rel. Grav.* 29:931, 1997.

Tomita, H. (1986). Curvature Invariants of Random Interface Generated by Gaussian Fields. *Progress of Theoretical Physics*, 76:952–955.

Tye, S. (2006). A New View of the Cosmic Landscape. *ArXiv High Energy Physics - Theory e-prints*.

Tye, S. and Wohns, D. (2009). Resonant Tunneling in Scalar Quantum Field Theory. *arXiv:hep-th/0910.1088*.

Vachaspati, T. and Trodden, M. (2000). Causality and cosmic inflation. *Phys. Rev. D*, 61(2):023502–+.

Vielva, P., Martínez-González, E., Barreiro, R. B., Sanz, J. L., and Cayón, L. (2004). Detection of Non-Gaussianity in the Wilkinson Microwave Anisotropy Probe First-Year Data Using Spherical Wavelets. *Astrophysical Journal*, 609:22–34.

Vilenkin, A. and Ford, L. H. (1982). Gravitational effects upon cosmological phase transitions. *Phys. Rev. D*, 26:1231–1241.

Wald, R. M. (1983). Asymptotic behavior of homogeneous cosmological models in the presence of a positive cosmological constant. *Phys. Rev. D*, 28:2118–2120.

Walker, A. G. (1935). On Riemannian spaces with spherical symmetry about a line, and the conditions for isotropy in general relativity. *Quart. J. Math. Oxford*, ser. 6:81.

Weinberg, S. (1989). The cosmological constant problem. *Reviews of Modern Physics*, 61:1–23.

Winitzki, S. (2008). Predictions in Eternal Inflation. In M. Lemoine, J. Martin, & P. Peter, editor, *Inflationary Cosmology*, volume 738 of *Lecture Notes in Physics, Berlin Springer Verlag*, pages 157–+.

Winitzki, S. (2009). Eternal Inflation. *ISBN:978-981-283-239-9. 236pp. World Scientific Publishing, 2009*.

Winitzki, S. and Kosowsky, A. (1998). Minkowski functional description of microwave background Gaussianity. *New Astronomy*, 3:75–100.

Yamamoto, K., Sasaki, M., and Tanaka, T. (1995). Large-Angle Cosmic Microwave Background Anisotropy in an Open Universe in the One-Bubble Inflationary Scenario. *Astrophysical Journal*, 455:412–+.

Zhang, R. and Huterer, D. (2010). Disks in the sky: A reassessment of the WMAP "cold spot" . *Astroparticle Physics*, 33:69–74.

Printed by Books on Demand GmbH, Norderstedt / Germany